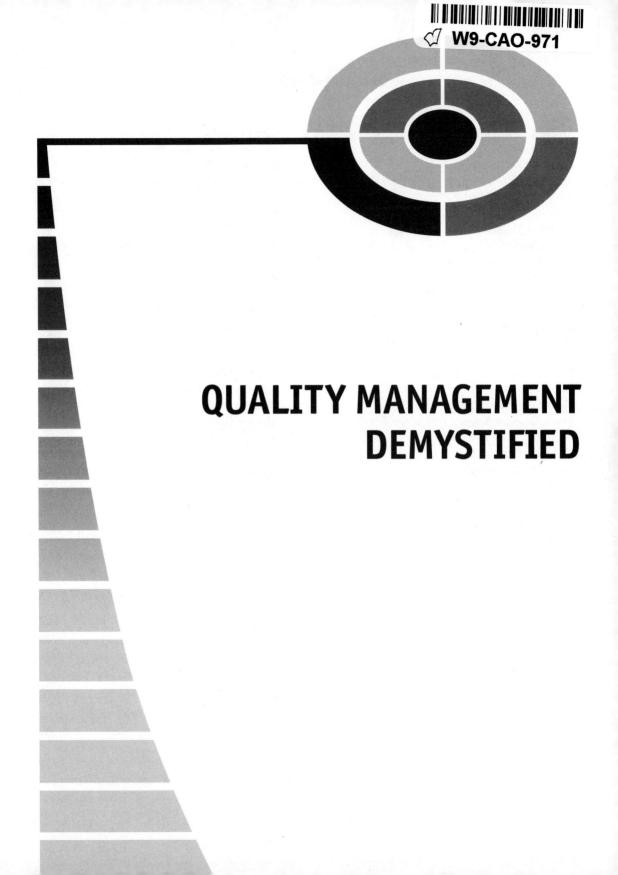

QUALITY MANAGEMENT
DEMYSTIFIED

Demystified Series

QUALITY MANAGEMENT DEMYSTIFIED

SID KEMP, PMP

McGRAW-HILL
New York   Chicago   San Francisco   Lisbon   London
Madrid   Mexico City   Milan   New Delhi   San Juan
Seoul   Singapore   Sydney   Toronto

The *McGraw-Hill* Companies

Cataloging-in-Publication Data is on file with the Library of Congress.

1 2 3 4 5 6 7 8 9 0   DOC/DOC   0 1 0 9 8 7 6 5

ISBN 0-07-144908-6

The sponsoring editor for this book was Judy Bass and the production supervisor was Pamela A. Pelton. It was set in Times Roman by American Color. The art director for the cover was Anthony Landi; the cover designer was Handel Low.

Printed and bound by RR Donnelley.

 This book is printed on recycled, acid-free paper containing a minimum of 50% recycled, de-inked fiber.

McGraw-Hill books are available at special quantity discounts to use as premiums and sales promotions, or for use in corporate training programs. For more information, please write to the Director of Special Sales, McGraw-Hill Professional, Two Penn Plaza, New York, NY 10121-2298. Or contact your local bookstore.

To my wife, Kris Lindbeck,
who sees beyond quality and value, to excellence.

CONTENTS

PREFACE

Many people, in many different ways, want to do good work. As people—in the business context, as customers—we all want to receive quality, to get good stuff, to get what we want. In the world of business—and outside it, in arts, hobbies, and personal growth—many of us want to deliver quality, to do good work, to deliver something of value to ourselves and others. *Quality Management Demystified* is about helping you do that in the context of business. This book is about how to get better and better at delivering value and doing good work, improving the quality of life for our customers, and the success of our own businesses.

I've spent many years trying to understand quality. And almost all of the books I've read open by acknowledging that quality is a difficult topic, that it is hard to define, that it is, well, mysterious. Most of the books either show one part of the puzzle—such as Quality Control or Six Sigma—very well and in depth, or catch most of the parts of the puzzle, but miss a few. Trying to do a jigsaw puzzle with missing pieces is frustrating. So, in writing *Quality Management Demystified* for you, I've done my best to give you the big picture and show you all the pieces. I hope to help you connect doing quality work, delivering quality results, and adding value to your customers and your company. All too often that connection gets lost. When the last piece of the quality puzzle dropped into place for me, the "aha!" moment really allowed me to improve my work and deliver better quality, and to help my customers solve quality problems, as well. I hope that you can use this book as a guide to solving quality problems and doing better at whatever you do.

Improving quality is possible in every type of work, from customer service to engineering to executive management. We have many reasons for wanting to do a good job—some want to excel, others to serve, and others to solve problems or make a bigger profit—and, in all of this, many of us have one thing in common: We strive for quality. Quality keeps customers by giving them the value they want; quality makes businesses succeed by delivering value; quality increases job satisfaction through our sense of accomplishment, of professionalism, and of service. People have been striving to understand, achieve, and deliver excellent quality for centuries. To put that another way, we all want quality, but we don't know how to get there, how to manage it.

Managing quality is an essential and difficult challenge. Quality is essential because it brings value. Without quality, we cannot bring value to our customers, we cannot realize value in our business, and we will soon be out of business and looking for another job. As important as quality is to business, managing to achieve it is extraordinarily difficult. Managing quality is as challenging as trying to manage life itself. Quality, in fact, is as large as life, and business has been struggling to bring quality under management for over 150 years. If we—as students, as workers, as managers, or executives—want to manage quality, then we do well to understand what quality is, and learn to achieve it. Managing quality is a gigantic job, and we cannot do it alone; we can only do it if we stand on the shoulders of giants. And there are giants—from the ancient Greek philosopher Plato, to Isaac Newton's student John Smeaton, who brought the scientific method to the work of engineering, to W. Edwards Deming, founder of the Total Quality Management movement—who have struggled to define quality and help companies achieve it. When we understand their efforts, we are ready to try to achieve quality in our own sphere—our jobs, our companies, and our lives.

The opposite of quality is error, and quality management is the effort to bring error under control and reduce error to acceptable levels. Yet, over and over, I see that most efforts at quality management do not do well. Efforts to reduce error are partial, incomplete, or abandoned too soon, before their value is realized. Usually, the main reason for these failures is that the people putting those efforts in place do not have a full understanding of quality. Quality seems much harder to define than on-time delivery, total productivity, or production at an acceptable cost. Compared to time cost and productivity, quality is a mystery.

The Parts of *Quality Management Demystified*

Enter *Quality Management Demystified.* Anyone concerned with quality can do better by learning from the successes of those who have come before. And that is what this book seeks to offer: a clarification of quality from the beginning of history to the twenty-first century, and an understanding of quality from its philosophical foundations to practical application in your business today. In Part 1: *Managing Quality* we learn how quality was understood and achieved by cultures around the world before it was defined as a part of engineering and business; how quality entered the business realm in the 1800s, and how, throughout the twentieth century, business has improved its understanding of quality and ability to manage quality in the workplace. We also understand how to make sense of the many different definitions of quality, and how to connect quality to our customers for business success. In Part 2: *Quality Essentials,* we learn the fundamental concepts and methods of quality management, including quality control, quality assurance, and more. We learn how to achieve quality as a human effort, through teamwork, and as an engineering effort, through the application of measurement, the scientific method, and statistics. Consistent quality improvement is only achieved through an effective combination of human and technical methods. In Part 3: *Quality Movements,* we will understand, evaluate, and compare methods around the world from 1950 to the present that have sought to bring quality and quality standards to business, industry, government, and education around the world. You will understand the strengths, weaknesses, perspectives, and benefits of each, so that you can choose a method, or improve the way you are using your current method. We open with Total Quality Management (TQM), the father of all the current movements. We then look at ISO 9000, the standard preferred in Europe; Six Sigma, the current approach in North America; CMMI, an attempt to bring quality into the field of software engineering; and *Gemba Kaizen* for just-in-time (JIT) manufacturing, the best practices of Japanese quality management.

All of this understanding is good, but doesn't really do a lot for us unless we can make it practical to our job today. As Native American medicine woman Dhyani Ywahoo says, when it comes to philosophy, the question is: Does it grow corn? In Part 4: *Practical Quality Management,* you will learn how to achieve real results in your business and your projects by applying a correct understanding of quality each day, week, and year.

To achieve quality, we must pierce its mysteries. Yet philosophers, engineers, and business management gurus have been struggling to do that for over 150 years. Don't let the scope scare you away; you don't need to become a philosopher, marketing expert, and process engineering guru all in one to achieve quality results. You only need to learn from those folks, not become them. I strive to keep the ideas simple and the style friendly.

How to Use This Book

Each reader comes to a book with a different purpose. In the case of *Quality Management Demystified,* we all want to do quality work, deliver quality results that meet specifications, and have a high-quality specification, so that the customer receiving the results gets value from the quality we deliver. Even with this common goal, our situations are quite different, and I have designed this book to be easy for readers coming from different perspectives:

- *The senior executive or business owner* can see how to organize a company or division to greatly increase effectiveness or efficiency, and to renew a quality environment that has been slipping. You will also be able to evaluate how well you use a method—such as Six Sigma, Total Quality Management, or Lean Manufacturing—to improve performance or select a new approach.
- *The team manager or project manager* can understand all the key ideas of quality, and see what is needed to make his or her team work.
- *The worker, technician, or engineer* can get the bigger picture of quality, and see ways of applying it to improve results on his or her job, make the job easier, and work more effectively with others.
- *The executive or manager in government, education, and not-for-profit work.* Although this book uses the language of business, the methods apply to all realms of productive activity and society: businesses of all sizes, government at all levels, not-for-profit organizations, and educational institutions.
- *The student of quality* can fill in crucial gaps in understanding quality that are left out of many texts, and test his or her understanding with quizzes at the end of each chapter and exams at the end of each part.

SIDEBARS FOR EASY LEARNING

As you move through *Quality Management Demystified*, you'll have lots of help. First of all, if you come across a word you don't know—or if you see that it's an ordinary word like "error" or "efficiency" has a technical meaning—check the glossary at the back of the book. What's more, we'll do all we can to help you keep your eye on the ball—the Q-ball to be exact. To play pool, you have to keep your eye on the cue ball. So, to learn about quality, we've given you a Q-ball. Each chapter has a variety of sidebars to help you out:

- *Eye on the Ball.* When you see this sidebar, you'll get the key point—the focus—of this section.
- *Quick Quality Tips.* The *Quick Quality Tips* in this book are quick, easy reminders for key quality ideas.
- *All the Angles.* Quality management applies equally to leadership, to business management, to project management, and to technical work. The *All the Angles* sidebars show you how to use a single technique at all

levels of work, or show you how to approach one problem from all of these perspectives.

- *Q-Pro.* If you're ready to work with the best, then learn the *Q-Pro* tips and bring quality to the highest level.
- *Mis-Q!* We all make mistakes sometimes, but do we learn from them? Even better, can we learn from the mistakes of others, and avoid the cost of making the same mistake ourselves. *Mis-Q!* sidebars give you the chance to do just that.
- *Align your Q.* We don't learn just by reading, we learn by thinking and making the ideas our own. To play pool, you have to plan your shot, and align your cue stick. To get better at quality management, you need to take the ideas, make them your own, and apply them to your own problems. The *Align Your Q* sidebars will get you thinking for yourself.
- *Q-Up.* Now it's time for you to take the Q-Ball into your own hands. *Q-Up* sidebars give you a chance to apply ideas from *Quality Management Demystified* to your own work.

Playing pool—or learning about quality—might make you hungry. You'll need some high-quality food—tasty and nutritious—to keep you going. *The Ham-and-Cheese Sandwich: Our Case Study* will help. You can learn everything you need to know about quality while practicing on a tasty snack. You can do more than enjoy these case studies. If you pull out a pad and pen, you can do some good thinking and learning on each one. My own answers to the case study questions are available on my company's web site at *www.qualitytechnology.com/QMD*.

In addition, each chapter ends with a *Q-Ball Quiz,* a quick multiple-choice test, so you can test your understanding of key ideas and terms. The book has two exams—a mid-term after Part 2 and a Final Exam at the end. Answers to the quizzes and exams are found at the back of the book.

PERSPECTIVES ON QUALITY

Most of you—my readers—already have some experience and opinion of quality management. In fact, the field can be polarized, with people holding such strong opinions that that there is more noise than listening. I approach all aspects of quality with an open mind, and I hope you will do the same. One approach to good dialogue is to realize that our own experience is only a small part of all that is happening. Here are some examples of what I mean.

- *Is Quality Assurance (QA) undervalued?* Many people on quality assurance teams will tell you that it is—and that this is a common or universal problem. I'll agree that QA is often undervalued and not given the support it needs. I'll also say that it isn't always that way, and that it doesn't have to be that way.

- *Is Six Sigma great, or is it a huge waste?* I know people who will argue strongly each way, and each is speaking the truth of his own experience. However, if we step back for a broader perspective, we see that sometimes Six Sigma works and does great things for a company; other times it ends up a total mess. The key is not to make absolute judgments, but instead to assess, to learn, to understand why it works or it doesn't, so that we can make it work—or keep it working—for our own organization.

Although we are not meeting face to face, as I write this book, I picture us having an open-minded, friendly conversation. Since we've learned in different times and places, there will be confusion. The terms of quality management are not all clear cut. They grew up at different times to solve different problems. What we already know can get in the way of understanding something new and learning more. As we put the big picture together, there will be some temporary misunderstandings. I hope that you will be open-minded and willing to examine your own perspective. I want to meet you where you are and carry the conversation forward.

In lecturing about quality around the country, I have found that there are many different perspectives and methods, and many different people with many things to offer. There are also some people who are sure they have "the" answer, as if there is just one answer. Some people are devoted to one school; others call that school a fad or even a fraud. Some think problems in quality management can't be solved; others think that they were solved a long time ago, and that we just need to use what is already known.

I want to demystify quality for you. And to do that, I'm going to need your help. Please take a moment to ask yourself, "Where do I stand in relation to understanding quality and quality management?" See if you are like these people I've met.

- *Totally New to Quality.* Maybe you are a student taking your first class on Quality Assurance or Quality Engineering. Or maybe you have just gotten a job where quality is a hot issue or quality certification is a job requirement. Come to *Quality Management Demystified* with an open mind. Let this book's big picture help you put the pieces from other books and people all together into one whole.
- *Confused About Quality.* Maybe you're a manager who's been told quality is important, but you're not sure how, or why. Maybe you're getting mixed messages: One day, "Do it right!" the next day, "Just get it done!" *Quality Management Demystified* will work best for you if you step back, open up

to the fact that these questions trouble a lot of people, and follow the book step by step.

- *A Quality Control (QC) Expert.* Within the technical field of QC, this book will probably not go into as much depth as you would like. However, you can learn the role of QC within the larger quality picture.
- *A Quality Assurance Expert.* QA is a difficult field, mostly because it does not receive enough organizational support. *Quality Management Demystified* will show you how to gain influence to improve the value of QA to your organization.
- *A Quality Auditor.* Auditing is a misunderstood and undervalued profession. Using approaches you will find here, you will be able to increase the business value of your audit services, and sell value-added auditing to your organization.
- *A Quality Engineer or Six Sigma Expert. Quality Management Demystified* will help you identify critical success factors for your organization's quality program.
- *A Department Manager, Project Manager, or Team Leader* can learn how to make quality work within your department or team, and then influence the rest of the organization.
- *An executive bringing quality management into your business, or thinking about it,* you will understand the value of a quality improvement program leading to an organization that can continuously improve quality, and learn how to implement the program and methods that lead to success.

Quality Management is sometimes a contentious field, with people defending their favorite schools or methods, or criticizing an approach that they have seen fail. Yet all approaches have something to offer, and all approaches sometimes fail in implementation. Writing in this contentious field, of course, opens one up to criticism. So, I thought I would give my critics—imaginary critics, at this point—a chance to ask me about why I wrote *Quality Management Demystified* the way that I did.

Interviewer: Sid, most books would open up with a definition of quality management, and then have chapters on Quality Control (QC) and Quality Assurance (QA). You don't talk about QC and QA until Chapter 6. Why such a long introduction?

Sid: I find most discussions of QC and QA mystifying. If we begin with QC and QA, we don't understand the problems that the people who were developing these methods were trying to solve, or the ideas they already knew. There is too much assumed, and that creates mystery. In the first five chapters, I try to connect the reader with the experience of quality and value, I trace the history of the practical effort to create quality before QC and QA, and I trace the history of ideas that came together to become quality management. With this background, the reader can sit right next to Shewhart as he defines Quality Control, knowing what he knew and facing what he faced. He or she can sit right next to Deming and see why a bigger picture of quality was needed, and how that grew into TQM. With the historical situation clearly set and the terms defined, QC and QA are no longer mysteries.

Interviewer: When you do introduce QC and QA, you include them as only two of five processes. Normally, QC and QA are seen as the two activities that we engage in to manage quality. Where did the other three come from?

Sid: I chose to give all five processes—Quality Definition, Quality Planning, Quality Control, Quality Assurance, and Quality Delivery—equal standing because, from a practical perspective, we need to do all of them if we want to deliver quality to the customer. If we do all five, we bring quality and error under management, and manage them from beginning to end. Quality Planning (QP) is recognized by both the International Organization for Standardization (ISO) and the Project Management Institute (PMI). Quality Definition is usually called requirements specification, or the voice of the customer, or scope definition. I hope to integrate scope—what we are making—with value and quality, and highlight the essential problem of finding out what the customer wants by putting this process—Quality Definition—first. We can only deliver quality if we know what quality is to the customer of our products, services, and projects. So, definition is the essential first step in bringing quality under management. The last process—Quality Delivery—is often called Customer Satisfaction or Customer Delight. It deserves attention because, unlike QC and QA, it brings quality all the way to the customer. When we learn to deliver quality, we can meet our customers' expectations as well as specifications. We can delight our customers. That leads to repeat business, referrals, and success. This five-step framework also allows us to compare and contrast the different schools of quality management in a single framework.

Interviewer: Speaking of different schools, what would you say to someone who said, "Six Sigma is a complete quality solution. Why discuss other types of quality management?"

Sid: I would say, "You might be right." But whenever Six Sigma—or any other method, such as a Zero Defects Initiative—is implemented well, it is

implemented well because all fourteen of Deming's key points of TQM are done right. And any effort that fails, fails because it misses one of those points. Quality Management is a system, and we have to cover all the bases to succeed. If we understand the history and the ideas—if there are no mysteries—then we can make any particular methodology succeed. If we misunderstand quality management, then any method is at risk of running into problems in implementation.

Interviewer: What would you say to someone who said, "Total Quality Management is dead?"

Sid: I would say that classical Newtonian physics died 100 years ago, with the arrival of the Theory of Relativity, yet 99% of all engineering problems can still be resolved by Newton's methods. Similarly, whether we change the name or not, when we look at processes, we find that everything is based on TQM. ISO 9000 is based on TQM. CMM is directly derived from a TQM effort, where Michael Fagan at IBM was guided by Dr. Joseph M. Juran, a TQM guru, to develop Software Inspection. That led to zero-defect software, which, when applied at NASA, led to CMM. As for Six Sigma, historically, it is an extension of TQM. TQM manufacturing set a goal of 3 sigma. When that was achieved, people pushed the envelope to 4 sigma, 5 sigma, and 6 sigma. GE made big press when they announced a 6 sigma initiative. Six Sigma may be new in terms of marketing, but when we look at the functional processes, Six Sigma has refined TQM, but hasn't added anything truly new. At least, that's what I've found in my research so far. If someone wants to show me a Six Sigma process that is not based in TQM, I'd love to see it and share it with others.

Interviewer: What would you say to a manager, executive, or business owner who said, "I want to improve quality, but I don't know where to start. Which standard should I apply? What should I shoot for?"

Sid: I would say, "Start where you are, then decide where you want to be. And set a goal of making a stronger business with a better bottom line." Management is a part of business, and the purpose of business is to stay in business and succeed. Outside business, quality can be a goal in itself: The artist or craftsperson can strive for beauty, the scholar for comprehension, the scientist for understanding, the party host for the enjoyment of his or her guests—without a focus on the bottom line or on meeting a delivery date. But, in business, success—staying in business by delivering on time and making money—is either a primary goal or a key requirement.

So fit your quality improvement effort to your business. Start where you are. If you aren't at 3 sigma, don't shoot for Six Sigma. Get defined processes in place before you try to improve your processes. Most importantly, lead your team—your whole company if you're at the top—to focus on truly understanding what the customer wants and specifying it, then delivering to that specification so that delivery of quality adds value for the customer. Change the way you work, then help your team change, and then move that outwards to other parts of the company, vendors, distributors, and customers.

Be very specific in assessing where you are and where you want to go. What are your quality goals, and what business value will you gain by achieving them? Will you save money, increase sales, retain customers? Create a project that takes your company from where it is to where you want it to be. Then focus your team on that project. That is what companies in Japan did under Deming's guidance in the 1950s. That is what Ford and Xerox did in the 1980s. Get teams who are excited about quality, teach them, then let them apply their understanding to their own work. Let them see and share in the benefits of that. That first project—even if it comes in late or is not perfect—gets the ball rolling. Then you have a team that is ready to apply the same method again and again—not only to quality problems, but to on-time delivery problems, to cost problems, to business planning problems, to customer service problems—until you are steadily serving all of your customers in all ways.

Dr. Masaaki Imai, the founder of the Kaizen Institute, says that today's standard is the worst possible way of doing any given job. Too often, we strive to meet standards. If we take the approach of *kaizen,* continuous improvement, then working to standard is a habit, and we are always asking: How can we make this standard even better?

That is the route of *Quality Management Demystified,* the route to customer delight, employee loyalty, and business success.

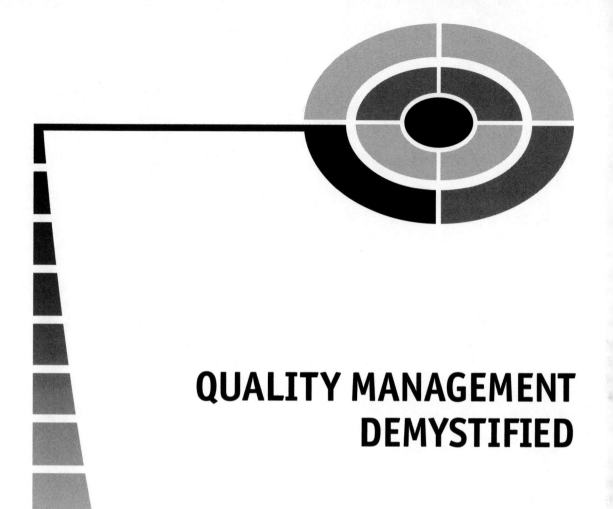

QUALITY MANAGEMENT
DEMYSTIFIED

PART ONE

Managing Quality

What is quality? What does it mean to manage anything, and to manage quality in particular? Our journey starts with an understanding of what quality is, a summary of centuries of thinking about quality, and a review of how business has tried to achieve, maintain, and deliver quality over the centuries. In Chapter 1: *Quality Throughout History,* you will see how quality has been a part of human life and civilization since the beginning of time. When we see how huge quality is, we see why it is so difficult to manage! Also, we will begin to explore the human experience—and our own individual experience—of the elements of quality. Chapter 2: *The Development of Quality Management* traces the history of the ideas that came together in the 1800s and 1900s to define the field we call quality management. We'll see how science began to change the way we design and build things, creating engineering out of the older traditions of craftsmanship. We'll see the problems faced by the industrial revolution and wartime that led to developing and using pieces of what later became quality management as components of the solution. And we'll see how quality management developed—sometimes in fits and starts—throughout the 20th century.

Many different scholarly disciplines have tried to define quality and help business achieve it, including philosophy, economics, marketing, and operations research. In Chapter 3: *Defining Quality,* we'll put all of those definitions together in a simple and practical way that will allow you to define quality in your business or on your project. We'll also see how we can get very practical if we see quality management as error management—the effort to bring errors and defects under control and reduce them to an acceptable level.

Once you've defined quality, you're on your way to managing it and delivering it. And, when we deliver quality, we deliver it to the customer. Chapter 4: *Quality for the Customer* addresses the first difficulty businesses face in quality management—figuring out what customers want. How do we get the customer to define what they want, so that we can deliver it? How can we avoid the common mistake of defining quality *for* the customer, instead of letting the customer define quality for us? If we can solve these problems, then we can hear the voice of the customer and get a specification of quality from the customer. When we deliver to that specification, the quality product or service we deliver adds value for the customer. If we can add value for our customers affordably, we add value to our own businesses at the same time.

By the end of Part 1: *Managing Quality,* you will understand the large job that quality managers face, and understand the problems that we have been working to solve for the last 150 years.

Quality Throughout History

In the past the man has been first; in the future the system must be first.
—Frederick Winslow Taylor, *Scientific Management,* 1911.

Quality is far larger than business. We talk about quality in all parts of our lives—in judging art, in evaluating the things that we make, in describing our experience. We even talk about quality time and quality relationships. In the broadest sense, quality is that which adds value, that which makes our lives better. Before we try to manage quality, we should try to understand our experience of quality. This chapter is dedicated to understanding the human experience of quality and to show how people have always strived to deliver quality, even long before quality management was defined.

Quality has been part of human life, culture, and history from its earliest beginnings, and it has always had two aspects. One aspect, represented by the 11,000-year-old Sphinx at Giza, is beauty. Indefinable and alluring, beauty draws us, adding richness to our lives. Another aspect is represented at Giza as well, the 5000-year-old Great Pyramid—still standing—represents the functional quality of great engineering. Both of these are shown in Figure 1-1. The Sphinx is still standing because definable and measurable functional quality brings stability to the more ephemeral quality of beauty. In recent centuries, we

Fig. 1-1. The Sphinx and Great Pyramid at Giza, Egypt,
photo courtesy of eStock Photo.

have been able to define more and more of what quality is, and, in defining it, make it more susceptible to engineering, make it reproducible, and bring it under management. But there will always be an indefinable side to quality—what we call beauty.

Quality Across All Cultures

Every culture in the world, on every inhabited continent—from the indigenous cultures that live closest to the earth to the highly evolved technical civilizations of the last 5000 years—has created great quality. The oldest works of living art come from the Australian aboriginal cultures. These nomadic tribes add to paintings each year, telling the story of the tribe. There is one painting that is over 20,000 years old, and still a work in progress. A little more of the tribe's story is added each year. Much of the world's beauty is ephemeral—arts that leave no trace such as music and dance, and those that fade quickly in time, such as basketry—so we do not have a trace of all the fine art, crafts, and engineering that have come and gone over the millennia. The work of indigenous cultures is often sophisticated and subtle, as complex and beautiful as anything created by more technically advanced civilizations.

Great civilizations have created larger, more enduring monuments in architecture and sculpture. Many of these have been inspired by religious traditions, including Hindu temples and statuary; Jewish temples and synagogues; pagan sites such as Stonehenge; Buddhist stupas—places for holy relics—temples, and statues; Christian cathedrals and statues; and Muslim mosques. The human awe at monumental structures and sense of inspiration from beautiful form is universal. It transcends all cultures.

Our sense of quality transcends time as well. Modern societies are impressed by skyscrapers and suspension bridges just as people of ancient cultures were amazed by monuments and temples. And we still seek to create beauty of form in our latest engineering feats. The materials, artistic media, and construction technology change, but the response to beauty and the value of durability, functionality, and other engineering qualities remain.

The Facets of Quality

What adds value? What is quality? Some philosophers, from Plato 2500 years ago in Greece, to Robert Pirsig in 1975, in *Zen and the Art of Motorcycle Maintenance,* argue that quality can't be defined, that we just know it when we see it. We'll discuss that more in Chapter 3: *Defining Quality.* For now, though, we can realize that something that brings us good feelings—especially feelings of rightness or goodness—or that brings healing, or that enriches us as individuals and as a society, are things that add value. All of these things have quality.

Philosophers can leave quality undefined, but those of us who want to succeed in business don't have that luxury. Our success is based on people who are our customers. Figuring out what people want—what people value—is essential so that we can deliver it to them and stay in business. You might say that businesspeople go where philosophers fear to tread—we want to understand what people mean by quality. Our definitions may not be philosophically sound, but they will be practically useful.

If we look more closely at the experience of quality, we see that there are four levels on which we determine that something has—or is lacking—quality:

- *Universal.* We are all awed by a night sky or a beautiful sunrise. All of us are relaxed by a swim in the sea and nourished by fresh water.
- *Cultural.* Some cultural value systems create agreements about what is beautiful that last for centuries. For example, to those who grow up learning

to appreciate classical music—whether it is of China, India, or Europe—that sound is wonderful. Yet the rules for each are different. Chinese music uses a five-tone scale, while Western music uses eight tones per octave. Chinese music and Western music always have rhythm, but Indian classical music includes a form that has no rhythm—no set beat or timing. To anyone trained in one tradition, the music is beautiful and deep. But the traditions alien to us, although alluring, are strange enough to be uncomfortable.

- *Social.* Many styles and forms are agreed on by groups of people—economic or social classes, ethnic groups, families, or groups of friends—that are smaller than a whole culture but larger than the individual. Styles may last a season, years, or even decades, but not centuries.
- *Personal.* Finally, there are some purely personal preferences regarding what matters, what has value, what is quality. As the French say, *chacun a son gout,* each to his own taste. Or, as the Romans put it, *de gustibus non disputandum est,* there is no disputing matters of taste. Or, in English, to each his own. So it is universally recognized that there is a valid personal element to taste, that there is no logical argument one can put forth to say what someone else should or should not like.

When we as people decide what we like, we make a decision internally, mixing together all four of the elements I just described. Usually, we are not aware of doing this. This mixing of levels is one of several things that make the customer definition of quality very difficult for people in business—people who want to deliver quality and value to other people. Here is a more complete list of issues that make it difficult to understand what people—our customers—want:

- *One problem ruins the whole experience.* If you go to a wonderful restaurant with excellent food, beautiful décor, and very slow service, what do you remember? The slow service. If a single part of the experience is not good, the whole experience falters, at least for most people most of the time.
- *Quality works at all four levels at once.* To experience quality, we must be satisfied at the universal, cultural, social, and individual levels. If the experience really fails for us on even just one level, it isn't an experience of quality.
- *Each individual is different, but companies succeed by selling to many individuals.* If we make all of our products the same, then we will completely satisfy a few people, but partially satisfy many others. That is the low-cost option. The higher-cost option is to provide custom products or services, where customers pay more to get exactly what they want, either

custom made or altered to customer specifications. Rather than two options, this is actually a range. In between standard identical products and full customization, we can offer products with options, or products with limited customization. The same applies to services.

As we work to define quality in the specific, we are asking, "What do customers like about our present or future products or services?" When we ask this question, we should remember the four facets of the experience of quality and keep in mind the complicated challenge of figuring out what other people—especially many other people who we don't know—want.

EYE ON THE BALL

Cultural Preferences Run Deep
My mother, a student of anthropology, heard this story from her teacher. The teacher was doing fieldwork with a tribe in Africa. The tribe had a custom of exchanging gifts of food. The anthropologist brought canned peaches in sugar syrup—a newly invented American delicacy—as his gift. At the ceremony, the chief gave his gift first—roasted ants. The anthropologist received the dish graciously, ate some, thanked his host, and then politely went behind a tree to throw up. He then offered the canned peaches. The tribal chief received them graciously, ate some, thanked his guest, and then politely went behind a tree to throw up. Neither one could stomach what the other thought of as a delicacy.

The lesson: Our assumptions about what everybody likes are often cultural, and not universal. And preferences about what we like are visceral and run deep. This is particularly true in relation to things that affect our sense of taste, touch, and sound.

Quality in Art and Engineering

Although we can talk about two aspects of quality—aesthetic or artistic, and functional or engineering—the distinction is not that simple. There are four key points:

- All art depends on engineering.
- Some art gets defined and becomes engineering.
- Changes in engineering create changes to art.
- Cultural values affect engineering methods.

ALL ART DEPENDS ON ENGINEERING

First of all, all art depends on engineering. That is, artistic quality—beauty—becomes unavailable unless it is supported by engineering qualities such as durability and stability. The best example I know is Vincent Van Gogh's masterpiece, *The Night Café* (*Café Terrace on the Place du Forum,* 1888). An oil painting requires a canvas prepared with a white surface called gesso. Gesso is spread on a stretched canvas as a thick, plaster-like paint. It must dry fully before the colorful oil paints are applied. In general, Van Gogh was a master of the technical aspects of his craft—canvas stretching, preparation with gesso, the mixing of paints. But this time, perhaps because he was excited and impatient to start painting, he didn't let the gesso dry. The result is that, over the decades, the painting has darkened. Now, no one can see the beauty of *The Night Café* as Van Gogh intended us to see it.

SOME ART GETS DEFINED AND BECOMES ENGINEERING

Many things that seem difficult or mysterious—that seem to be the result of creative genius to the outsider—are actually mechanical and obvious to the practitioner. Artistic performance—such as acting or singing—can appear to be "pure art," but the practitioners will tell you that a great deal of technique—technical skill that comes with defined practice—is the basis for that art. Perhaps this is most clear in the art of illusion—the performance of magic. If we don't know the technique, the engineering mechanisms, and the long practice involved, we see something astounding. But the practitioner knows how the trick works, practices over and over, and then adds a bit of flair—maybe 80% technique, 15% showmanship, and only 5% creative genius.

Similarly, some things that were once art become engineering. The artistry of mixing pigments to make paint is now a technical science. The characteristics of finely crafted musical instruments have now been evaluated and reproduced so that electronic chips can create sound that previously could only be made by a full symphony orchestra. Fractal images—computer-generated mathematics—can create images of whole planets of realistic mountains, forests, clouds, and oceans. Until recently, we knew that every snowflake was unique, but we didn't know how snowflakes are made. Now, we have a working scientific model of how unique snowflakes are made, and we can reproduce snow in the laboratory.

Art Becomes Engineering

When nature and art are analyzed, we learn how natural and artistic results are created. When we understand this well enough, that which was unique and creative becomes reproducible. The unique and creative—art—is analyzed through investigation—science—resulting in repeatable processes—engineering.

CHANGES IN ENGINEERING CREATE CHANGES TO ART

In many ways, changes in engineering and technology change art:

- New technology creates the possibility of new art forms. Think of laser light shows or virtual reality movies and experiences.
- When what was an irreproducible artistic technique becomes a reproducible engineering technique, the art form becomes more available and less expensive. However, work in the medium also becomes easier. With more people able to produce in the medium with less time spent working and practicing, there is more poor quality art produced in the medium. There is also more innovation and excellence produced, as well.
- When engineering makes something that was rare and expensive common and cheap, our valuation—our idea of value and quality—changes. Before the 1880s, aluminum was a rare and exotic metal used only for fine jewelry. When Charles Martin Hall invented an inexpensive process of extracting metallic aluminum from bauxite, a plentiful rock, aluminum became the cheapest and most readily available of all metals. It is no longer seen as rare or valuable and is rarely used in jewelry.

THE HIGHEST QUALITY REQUIRES BOTH ART AND ENGINEERING

As we've seen, we cannot completely separate art—the aesthetic, indefinable, irreproducible side of quality—from engineering—the technical, more definable, more measurable, more reproducible side of quality. We can appreciate both. We can appreciate ephemeral—short-lived—beauty, and we can also appreciate the durability of something that is useful even if it is ugly. But when architecture or sculpture is both beautiful—offering the highest level of artistic or aesthetic value—and also impressive for being monumental in size, durable over the years, precise in the way its form fulfills its function, it is greatly admired for generations.

We can understand this topic by thinking about two different meanings of the word *design*. In engineering, design is about intelligent problem solving. A good engineering design solves a problem effectively at lower cost than previous designs. But, in relation to aesthetics, design is about style, and is independent of solving any technical problem. It is about aesthetic appeal, and relates to the social level of aesthetic quality. Most design involves compromise—either a compromise of getting less of one function for more of another, or a compromise between functionality and beauty. Great design is not a compromise, it is a solution that offers both beauty and functionality.

CULTURAL VALUES AFFECT ENGINEERING METHODS

We don't realize it, but we are always thinking inside boxes, inside a limited framework. Thinking outside the box is a useful exercise that can give us better solutions, but usually, when we jump out of one box, we jump into another—but perhaps a bigger box with more options. To think outside the box means to identify our assumptions and question them. This is a very good method in quality engineering because it can provide more effective, less expensive solutions, or even solutions to problems that seemed unsolvable.

One box we are inside is a cultural box, and we only see that we are in it if we start to study other cultures in depth. Once we understand another culture, we can see that our own culture's values and concepts are cultural and relative, and not absolute. One example is particularly relevant to quality engineering: As we discussed above, there is a universal appreciation for that which lasts a long time. When I ask you to think of a building or monument that lasts a long time, what do you think of? If you are a Westerner, like myself, you probably think of something made of stone or metal, something durable. In the West, we tend to think of longevity and durability as being one and the same.

Not so in Japan. In traditional Japan, longevity was achieved by renewal, not by durability. Partly, this was due to their appreciation of beauty as transient and ephemeral. But it also reflected a different perspective on how to make things that last a long time. This is perhaps illustrated best by the Shinto temple at Ise, a shrine that is over 1000 years old. But it is not durable. Instead, it is renewable. It has been rebuilt every 20 years for over 1000 years. Instead of making the building last, the Japanese maintained the craft skills and engineering designs and methods, and rebuilt the temple more than 50 times.

In the West, we wouldn't think of making something last by tearing it down and building it over and over again. That is because we associate the universal value—longevity—with the cultural value—durability. When we see that another culture doesn't have the same association, we can see our own cultural framework, the box that we think in.

In looking at many examples of beauty and quality engineering across all cultures, we begin to separate the four levels of thinking and feeling that make up our sense of value and quality. We move beyond our immediate response of knowing either that we like something or we don't. We move towards understanding the elements of quality, towards defining quality.

And defining quality is the first step in bringing it under management.

Quality Before Business

If people were creating quality across the world for thousands of years before the invention of quality management, how did they do it? What can we learn from them? The two oldest ideas that became part of quality management are the idea of a standard and standardization and writings and schools. Let's take a look at standards and schools, and how they developed. Then we will look at another tool used to sustain and transmit quality—secret teachings.

STANDARDIZATION IN LAW AND MEDICINE

A standard is a rule or guideline that, when followed, brings consistency. In relation to business, we can identify standards in two broad categories: external and internal. External standards came around a long time before internal standards.

External Standards

External standards are those rules that keep the business environment—things external to business—stable. Most external standards are either customs or laws. A stable society can rely on relatively unchanging customs—rules of negotiating and keeping agreements, general codes of business conduct, and so forth. But when times become difficult—with famine, war, or different cultures mix— then cultural standards become unreliable. In these times, if the rule of law can be maintained, then businesses can survive and perhaps thrive. If the rule of law is lost, then business becomes very difficult, and often shady or criminal. So law and stability are valuable for business.

Many times throughout history, some people have been above the law. These people—usually royalty or the most wealthy classes—could do as they pleased, whether there were any laws or not. The first time in history that changed was with the Code of Hammurabi, in 1780 B.C.E. This was a written legal code that even the King had to obey, and it was a huge step forward in civilization and stability. When people know the rules, and the rules are written down and change only slowly, we can learn how to do business and count on being able to do business in the same way year after year. We can get better at what we do. The stability of the rule of law, and the elimination of arbitrary authority or advantage for those who are above the law, makes room for improving quality, effectiveness, and efficiency, rather than always trying to cope with changing circumstances or trying to gain favor with those in power.

REGULATIONS

The first laws governed crime and also how law would be administered. Later, rules for businesses were developed. This was the beginning of a special kind of external standard, a regulation. A regulation is a standard with the power of law. If we violate a regulation, then the company is breaking the law, and there are penalties imposed by the government. Some regulations are stated in law. Other regulations are backed by law but stated in documents produced by the executive branch of government and published as regulations.

Since regulations tell all businesses how to behave, they are both external and internal standards. Regulations make things regular; they put us all on a level playing field. From the perspective of a single business, they regulate both the external environment—the activity of other businesses—and internal activity— how we operate inside the company. Regulations keep all the players—all the competing businesses—running by the same rules. They also set rules that make things fair for customers and businesses when customers buy from businesses.

External standards and regulations are important to business, but they are not a central part of quality management. Still, we can't ignore them. One of the responsibilities of quality management is to ensure that all business processes and activities comply with regulations. And we can improve the quality of what we offer in a given culture and country by understanding its laws and customs. In our increasingly global society, quality management often means understanding international and multicultural laws, customs, and values.

MIS-Q

Who Cut the Hoses?

When fire companies first developed, they were not run by cities. According to one story, they were run by insurance companies. A business owner would buy fire insurance from a company, and then that company would send a fire truck if your business caught fire. This was very inefficient: When your business caught fire, you would have to wait until your company came, even if another company was nearer.

Worse, other fire companies would come and cut the hoses of the company trying to put out the fire! They figured that if they could ruin the other fire insurance company, they'd get more business.

Clearly, when it comes to putting out fires, an unregulated business is a bad idea. Society solves this type of problem by either making the industry public through city and volunteer fire companies or by regulating the industry, as are electric and gas companies. Public companies and regulated industries have their problems, but they are not as bad as what came before.

The lesson: Regulations are put in place to solve problems of unfairness and social chaos. They sometimes create new problems as well.

Internal Standards

Although external standards—in the form of laws—have been around for thousands of years, internal standards mostly came later. They mostly took the form of trade secrets—formulas, recipes, and methods kept in secret, perhaps written down, or perhaps simply taught from parent to child or master to apprentice from one generation to the next.

The benefit of internal standards is consistency. And, if you know how to deliver high quality, then, through a standard, you can deliver consistent high quality. When we repeat the same process using the same ingredients, we tend to get the same results. If one person does that well, we call it craftsmanship. When one person passes it on to others who seek to maintain it, it becomes a standard.

Standards are an important part of quality today. Departments and companies have internal standards, industries have standards, and independent or government-supported agencies offer standards, as well. The difference between a standard and a regulation is that there is no law that says that you have to follow a standard. It's just usually good business sense to do it. For example, if you want to make a device that runs on 150 volts instead of the standard 120 volts of household current in North America—or 220 volts for Europe—you can. But nobody could use your product, because, if they plug it in, it won't work.

Determining what standards your product or service should meet, and then assuring that you do meet those standards, is an important part of quality management. We'll discuss the development of standards more thoroughly in Chapter 2: *The Development of Quality Management.*

WRITINGS AND SCHOOLS

In ancient Greece, ancient China, and ancient India, the earliest compilations of written standards were textbooks on medical treatment. Herbal remedies, acupuncture, and other treatments have been specified for thousands of years. Medicine was probably the first standardized profession because of its importance. Many medicines, if given in the wrong dose or to the wrong patient, can kill instead of cure. So passing on careful instructions about what works is very important. Written instructions—formulas or recipes—were an easy way to remember diagnostic methods, remedies, and treatments and to pass them from place to place and generation to generation. Of course, it was also important to identify sources of medicines, so books to identify herbs and plants were valuable.

Medicine, law, philosophy, and religion made up a large bulk of the written material in the world before the invention of the printing press with moveable type in 1500. Books were rare, and so were people who could read. Although we now think of books and writing as a way to share information, in ancient times, the written word was often a way to keep something secret. This opens up our next topic: secret teachings.

SECRET TEACHINGS

Secret teachings were probably the main way that production methods and quality methods were passed down from generation to generation before the 1800s. Each craft—all the metal smiths, the makers of weapons, potters, bakers, and many others—had its own guild. The guild was led by masters. Masters would take on young apprentices. When an apprentice had worked long enough and learned enough skill, he became a journeyman. After many more years, by proving his skill, he became a master.

Guilds had standards of conduct for how they did business, and how they plied their craft. In the best of times, guilds helped young men become mature members of society and contributed to the well being of the community by doing good works of service. We see remnants of these traditions in organizations like the Lion's Club and the Shriners. You might note how many Shriners hospitals there are around the country.

Guilds and similar organizations in many countries provided cultural stability, improved the general well being of society, and also maintained quality. They did so by passing down a way of living—and specific methods—from one generation to the next. But these organizations did not separate out their moral rules and social customs from quality methods. It was all part of a way of life. Also, bad methods could be passed along as well as good ones. If something has been passed down as a secret teaching, revered for generations, it is easy to cling to it and ignore or repress a better method that someone might discover.

Guilds were not the only holders of secret teachings, nor the only organizations that could pass on quality for generations and centuries. Along with secular guilds, there were also religious orders. Monastic orders created and maintained methods of making cheese and wine in Europe, and elaborate cooking and the martial arts in China and Japan. Whatever the community, in writing, or by means of training from master to apprentice or disciple, ancient cultures passed on methods of creating beauty and lasting value. Repeatable process—the foundation of quality management—has been with us for centuries.

Ancient Quality—Maintaining, But Rarely Improving

You may have noticed some things missing in this discussion. We have talked about how quality was maintained in ancient times, but not how was it created in the first place, or improved. How was quality first created or improved in ancient times? The answer comes down to one word: genius. The particular creativity, innovative skill, drive, and talent of a single individual or an inspiring teacher or leader would lead to improvements. If those improvements were accepted, they were passed along. But there was no standard way of making improvements happen. That didn't come along until the scientific revolution, just a few hundred years ago.

When things don't improve, they tend to deteriorate. Before the scientific revolution, there were two opposing forces acting on the quality of the products and services created for society. Genius, plus people's desire for quality—the desire to get more value—tended to make things better. Human fallibility—the ability to make errors—and the lack of response to change, tended to make things worse. If someone solved a problem and the solution was accepted, things got better. If a good method was forgotten, or circumstances changed but people kept working in the same old way, things got worse. And there were no rules or methods that showed that one thing was definably, measurably better than another. So, if two people both came up with a change, the better change might not be accepted. Maybe the person with the change that wasn't as good was a friend of the king. When things got better or worse, it was often the luck of the draw.

Q-UP

Is Your Company Following the Ancient Way?
Even today, many companies only improve through genius, and through a hit-or-miss political process of acceptance of new ideas. What's true for your company? Do better methods get adopted? When and how? Is there a defined process for proposing an improvement, demonstrating it is better, and then putting it in place? If not, then your company is missing out on almost 200 years of good business practice. Keep reading; Quality Management Demystified will be a big help.

In spite of this, we can observe some quality principles at work. Certain things work well due to unchanging principles of nature. For example, all roads—whatever they are made of—are worn by rain pouring down and running off them, by the effects of water and gravity. As a result, the way to make a durable road will always be a way of building a road that handles water well. The ancient Romans discovered this, and there are Roman roads over 1400 years old that you can still walk today. The Roman roads were built in four layers. The bottom had large stones. Above that, there are broken stones: Pebbles and sand were mixed with cement to form a strong base. The third layer was made of cement and broken tile. The top was carved paving stones, laid closely together to create a flat surface. At the side, curb stones held the paving stones in place and gave a channel for water to run off.

Today, a good road is made in almost the same way. The modern road was invented by John Louis MacAdam in the early 1800s, and it bears his name: macadam. The lowest layers are made of large, broken stones. Then there is a layer of finer stone. The road is sloped and ditches are added at the side to provide drainage. The only change to the roads since the early 1800s is that, when cars were invented, they went fast enough to pull dust up from the road. This was solved by spraying the road with tar, creating today's black roads made of macadam with tar, or tarmac.

At first, it seems surprising that the ancient Roman roads are so similar to today's superhighways. But, on reflection, it makes a lot of sense. Most of the problem—to create a flat surface for wheeled vehicles that would last a long time—hasn't changed. As a result, most of the solution hasn't changed, either.

We use a mix of larger and smaller stones in layers, and shape the road to let rainwater run off. The one new problem—air suction from fast vehicles—was solved by a new solution: tar. As we look at quality engineering and quality business processes throughout this book, we will see this over and over again. Lasting principles lead to similar methods to achieve good results, but the best application of those principles changes in different situations.

EYE ON THE BALL

Universal Principles Applied in Changing Circumstances
High quality is always achieved by an understanding of universal, unchanging principles combined with a practical application of those unchanging principles to changing circumstances. This is as true of quality in method—business practices—as it is of quality in technique—artistic or engineering practices.

Conclusion:
Quality All Around the World

Quality has been with us for as long as there has been life on Earth, and people have been striving to create quality for as long as we have been around. Until a couple of hundred years ago, the creation of quality relied on individual expertise, transmission through a few standardized texts and schools in some specialties, and transmission through folklore and secret teachings. As a result, quality was very much a hit-or-miss affair. In Chapter 2, we will look at the history of how we have moved, in 200 years, from this situation to a situation where we can investigate and define quality, measure quality, repeat processes to deliver consistent quality, and improve quality in business.

ALIGN YOUR Q

Ham and Cheese Around the World
Think about these questions to see how culture affects quality, and how business managers have to think to provide quality to customers.

1. How many different combinations of ham and cheese can you think of? What country or culture does each one come from?

2. Pick another food that you know—sushi, or peanuts, or anything else—and describe the different dishes and tastes, and how different cultures have changed them.
3. You have been asked to manage catering for a multicultural, inter-religious event. Your caterer recommends ham and cheese. You check with your customers, including traditional Muslims and Orthodox Jews. You learn that the first don't eat ham. The second eat neither ham nor mix milk and meat—and they won't even eat food prepared in a kitchen where ham has been prepared. You also learn that several of the guests have already pre-ordered their meals, including ham-and-cheese sandwiches. Come up with three to five solutions, and choose the best one if price is not an issue, and the best one if you have to keep costs down.

And yet many companies are still doing things the way that they were done for thousands of years. When a genius does that, it works—because geniuses have always been able to deliver quality. But when average people do it, the results are—no surprise—average. If average people want to deliver high quality—or if geniuses want to deliver excellence over and over—we need to know how to be consistent and improve. We need quality management.

Q-Ball Quiz for Chapter 1

1. The two main dimensions of quality are
   (a) ancient and modern.
   (b) aesthetic and functional.
   (c) external and internal.
   (d) functional and engineering.

2. Which of these is *not* one of the four levels of the experience of quality?
   (a) social
   (b) universal
   (c) religious
   (d) cultural
   (e) personal

3. Which of these is the best question to ask when defining quality in business?
   (a) What is best for the company and its customers?
   (b) How can we improve our products or services?
   (c) How can we add the greatest value?
   (d) What do customers like about our present or future products and services?

4. All things deteriorate over time. Which of these is *not* a reason for the deterioration of quality?
   (a) Standards change.
   (b) Things wear out.
   (c) The situation changes, but people keep doing the same thing.
   (d) Students sometimes learn poorly and don't do as well as their teachers.

5. Which of these is *not* part of the way quality was passed down in ancient times?
   (a) written texts
   (b) secret teachings
   (c) the scientific method
   (d) apprenticeship in guilds

CHAPTER 2

The Development of Quality Management

To try to convince the reader that the remedy for this inefficiency lies in systematic management, rather than in searching for some unusual or extraordinary man.
—Frederick Winslow Taylor, *Scientific Management*, 1911.

The scientific revolution (roughly 1600 to 1687), which changed our perception of the universe, was followed by the industrial revolution (roughly 1760 to 1830, and beyond), which changed the way we live our lives. The relationship between science and industry—particularly the relationship between the scientific method and business management—gave rise to quality management. Quality management evolved as engineers, managers, executives, and government officials responded to the problems of their day. It was an evolution, not a planned series of developments. It is fair to say that, as we look back in history, some people and events seem more important to us now than they were thought to be at the time, and others seem less important now than they seemed then. I would argue that this understanding of quality management is still evolving and incomplete.

EYE ON THE BALL

The Evolution of Quality Management
The key to understanding quality management is to understand how people, solving problems generation after generation, slowly saw new ways to apply the scientific method to engineering and business, and also began to separate business management from technical engineering.

Key Ideas

The two most essential ideas at the core of quality management are standardization and the scientific method.

STANDARDIZATION

In Chapter 1: *Quality Throughout History,* we introduced the idea of standards. Now, we're going to take that one step further, to the idea of standardization. A standard is a defined, observable, measurable target, goal, or requirement to which we try to conform. Standardization is the process of finding out if we are conforming to the standard and then correcting what we do, so that we conform more closely to the standard.

To standardize something, we have to:

- Understand the standard.
- Have a way of comparing the thing or process to the standard.
- Know how much variation from the standard is acceptable.
- Take action when items do not meet the standard—throw them away or fix them, and perhaps fix the process that created them.

If we take these steps, then we will be able to make our processes and products meet standards more and more. In this chapter, we will see how

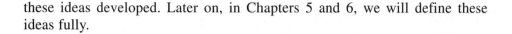

these ideas developed. Later on, in Chapters 5 and 6, we will define these ideas fully.

EMPIRICISM, MATHEMATICS, AND THE SCIENTIFIC METHOD

One of the key defining elements of science is empiricism—that is, the idea that observable facts and experience are the basis, or the most important basis, of knowledge. This distinguishes science from some religions and philosophies that grant authority for truth to a divine source or to a book from a divine source. Science looks at what is, and business needs to know what works. A small logical step from empiricism leads to a practical aspect of engineering and business: Doing what works and not doing what doesn't work is good for business.

Also, science frequently uses mathematics in many different ways. As we shall see, quality management has picked up the use of mathematics—especially statistics—from science.

A third key aspect of science is the scientific method, which is the process that scientists use to make and test theories. Here are the steps of the scientific method.

1. *Observe* something in nature or reality.
2. *Create a hypothesis,* a statement of a possible reason why the observed event happens.
3. *Design a test,* that will give observable results that can be interpreted to evaluate the hypothesis. For example, a test might show, "If X happens, the hypothesis seems to be true, but if Y happens, the hypothesis cannot be true." A test might also be designed to show which of two hypotheses is true.
4. *Perform the test and record the results.*
5. *Evaluate the results of the test.* If a test confirms a hypothesis, then it moves towards being a theory, which is simply an accepted hypothesis.

By applying the scientific method repeatedly, scientists seek to understand how the world works. The various fields of science apply this to the physical world in physics, astronomy, and chemistry; to life in the biological sciences; and to humanity in psychology and the social sciences.

Let's take a look at the history of the development of quality management. As you will see in the following sections, three of the most significant moments in the development of quality management occurred when the scientific method was applied. That is, not to understanding how the world works, but to understanding how to make technology and business work better.

Smeaton and Scientific Engineering

John Smeaton, an English physicist inspired by Newton, was the first person to bring the scientific method to the world of engineering. In 1755, he took on the job of building the third Eddystone lighthouse on the coast of England. The first two lighthouses had been destroyed, and Smeaton took on the task of building a lighthouse that would last. His key tool was the idea of experimentation. He succeeded. In fact, Smeaton succeeded so well at building things that he became known as the father of civil engineering. And behind every one of his successes and every one of his innovations was one method: the application of the scientific method to engineering problems. That is, Smeaton followed the steps of observation, hypothesis, and testing, not to produce theories about how the world works, but to produce practical methods that make things that work.

His remarkable accomplishments include:

- A recipe for concrete called hydraulic lime, the strongest to date, and the strongest for over another 100 years.
- Dovetail fitting of granite blocks, which he showed gave maximum strength in construction.
- Specialized cranes used for construction projects.

In all the thousands of years before John Smeaton, people built towers and bridges and buildings and then hoped they would stand up. Smeaton changed all that by testing what would stand and what would fall, that is, by performing experiments and then using the results of those experiments to figure out the best way to build anything. By the time he built the structure, he knew that it would work, because its components and processes were already tested.

A key element of the scientific method is that the experiment and its results are written down precisely. That allows others to repeat the experiments so that they can be verified, or so that they can be challenged. In business, we find that writing down our methods is essential to quality management. On the technical side, we need engineering specifications. On the business side, we need standard operating procedures. If we don't have these, then our work will tend to deteriorate. If we have them, we can use them as standards and then keep improving the way we work.

The Eddystone lighthouse itself is still standing in a field in Devon, as you can see in Figure 2-1. It was moved inland as a historical monument when it outlasted the rocks it was built on. It's beauty and durability shows John Smeaton's genius: Not only did his work encompass the technical, engineering side of

Fig. 2-1.   Smeaton's lighthouse, illustration by Kris Lindbeck.

quality, it also used the aesthetic side. The shape of the lighthouse was modeled after the trunk of an oak tree. Smeaton didn't know why oak trees stand for so long, but he knew that they did, and he copied the shape in building the lighthouse instead of using a simple conic section with straight sides. In imitating nature, he increased both the durability and, in my opinion, the beauty of the lighthouse.

Industrial Standardization in the 1800s

The next major step in the development of quality management had to do not with science, but with standardization. As early as the 1700s, mass production began through early versions of the assembly line, where each worker would make one part of an item, and then the item would be assembled and finished. The earliest items made this way were probably guns. Certainly, the history of guns, munitions, and war materiel was crucial to the development of quality management. There is a very simple, practical reason for this. In wartime, large quantities of equipment need to be delivered on time, and every single one needs to work. In peacetime, delays are costly; in wartime they are deadly. In peacetime, delivering bad quality means losing a customer; in wartime, delivering bad quality means the customer may lose his life.

INDUSTRY STANDARDS

As we discussed in Chapter 1: *Quality Throughout History,* external and internal standards have existed for centuries. In the early to middle 1800s, in time for the Civil War, a new type of standard arose: the industrial standard. This standard was the solution to a problem faced by the U.S. military.

Guns are built of multiple parts that must be fitted very closely together so that there is no leakage of pressure or burning material when the gunpowder explodes. There can also be no blockage of the bullet's path down the barrel and out of the front of the gun. Early assembly lines were already being used, but, prior to the Civil War, each gun was assembled by hand, with parts filed down to a smooth fit if necessary. As a result, each gun worked, but it was unique. Only a gunsmith with the proper training and skill could assemble a gun, and it could only be done at the factory.

But guns break and are damaged in the field of battle. The army wanted guns with replaceable parts. That way, if the barrel of one gun was damaged, and the cylinder of another was damaged, the working parts of two guns could be reassembled to make one working gun. Or spare parts could be brought to the battlefield, so that guns could be fixed. In addition to replaceable parts, the army also wanted two suppliers of each type of gun. That way, if one supplier was cut off by the enemy—either by destruction of the factory, by cutting of supply lines, or by defection—the army could get the guns it needed from somewhere else.

For the army to get what it wanted, two different companies would have to produce parts of guns so similar that they could be mixed at the field of battle and assembled to make a working weapon. That required the customer—the army—to define a specification for its rifles and pistols. It also required every manufacturer in the industry to meet that specification for each part that it delivered. Since the entire industry—gun manufacturing—had to meet one specification, it was called an industry specification.

Meeting the army's needs required one new idea that is crucial to quality management: the notion of tolerances. It was no longer good enough to say, "the cylinder of the gun must be exactly 1.5 inches long." What does "exactly" mean? Can we be off by a tenth of an inch? A hundredth? A thousandth? A *tolerance* defines that exactly. For example, we might say, "The cylinder of the gun must be 1.5 inches long, plus or minus 0.05 inches." Or we might have a closer tolerance, "The cylinder of the gun must be 1.5 inches long, plus or minus 0.01 inches."

If each manufacturer made each part to specifications and within tolerances, and made sure that the part met that industry standard, then the army could be confident that all the different pieces of a gun could be replaced in the field—even by parts from different manufacturers. In solving this problem of weapons supply, the army developed three key concepts that would be part of quality engineering: industry standards; tolerances; and inspection of every part to make sure that it met the industry standard, also called a specification.

MIS-Q

100 Years Behind in Quality Management
In the early 1950s, the first several thousand Xerox photocopiers were built and delivered to customers. When they broke down, the company would send field representatives with spare parts to replace the defective or broken part. The company found very quickly that the spare parts often wouldn't fit. Among other problems, the machine's cabinet was being built by a cabinetmaker who still used hand finishing to fit parts together. The company defined tighter tolerances that would allow for field-replaceable parts, employing a solution first developed 100 years earlier for the military.

The lesson: Some industries often lag far behind the best practices developed for the military, aerospace, and medicine, where lives are at risk. What quality management techniques could bring your company up to date?

THE SECOND INDUSTRIAL REVOLUTION

The second industrial revolution is generally dated from 1871–1914. It is marked by the development of modern printing and electronic communications and the creation of smaller, more powerful engines that transformed transportation, particularly with cars and airplanes beginning to replace steam locomotives and steamships. What is not generally realized is how key elements of what would later be quality management were essential to the second industrial revolution. Application of the scientific method was a major source of improvements in every industry, because it became generally accepted that one could prove, by means of empirical comparison or experiment, that one machine was better than another, or that one way of working was better than another. Without realizing it—in fact, the idea would not be defined for another 100 years—we had begun to use the idea of feedback, of using information to adjust and improve the way we work. While Smeaton was introducing feedback into the human process of design, automated feedback for automated machinery was developed because it was essential to the safe and reliable machine operations. In 1788, James Watt introduced the centrifugal governor—a device that regulates the power of an engine—to the steam engine. A governor keeps an engine running steadily even under varying loads. A modern example of a governor is a car with cruise control, which will maintain the same speed even if the car starts to go uphill, because more gas is sent to the engine to maintain the same speed. If the car starts to go downhill, then it speeds up due to gravity, and the cruise control detects that and sends less gas to the engine, keeping the car at a steady speed.

The notions of governance, regulation, control, and feedback are all essential to quality management in development of business, communications, and technical processes, and in the automated control of equipment and production. These ideas were essential to the second industrial revolution and in wide use, but no one had yet realized that they were all related ideas, or that they applied to business processes, to human resource processes, and to engineering.

INSPECTION

By the late 1800s, many things were being made in factories, using powerful furnaces and engines, powered tools, assembly lines, and inspection. At the end of each assembly line, every single component or product was inspected to make sure that it either met specifications or that it worked. Items that didn't work were either discarded or reworked. A discarded item might be thrown away, or it might be broken down or melted down into components that could be reused. The choice of discarding or reworking items depended on cost—whichever was less expensive was done.

Taylor Introduces Scientific Management

The biggest change for the worker from the pre-industrial period of craft guilds to the industrial revolution was the breaking up of work into tasks that were then ordered into an assembly line. The division of tasks was thought out, sometimes carefully. However, the method of doing the tasks was not. This was discovered by Frederick Winslow Taylor, who spent a lifetime developing and applying the principles of *scientific management,* and published his results in the treatise "The Principles of Scientific Management" in 1911. Scientific management is the direct predecessor of all of quality management. Indeed, a close look at Taylor's treatise shows that we have not yet consistently implemented all of his ideas, and that most failures in quality management over the past 100 years can be traced to errors in the application of management methods—errors that he warned against.

Taylor's key ideas can be summarized in two points:

- The scientific method can be used to define, and then to continuously improve, the best tools and methods for doing any job.
- Changing an organization's way of working requires principled management that makes the situation better for both management and workers. A slow, careful method must be followed so that management and workers work together towards common goals, and conflict is eliminated, rather than triggered, by the change in work process.

No one can explain these two points better than Taylor himself, in "The Principles of Scientific Management," Chapter 2:

"The general steps to be taken...are as follows:

- *First.* Find, say, 10 or 15 different men (preferably in as many separate establishments and different parts of the country) who are especially skillful in doing the particular work to be analyzed.
- *Second.* Study the exact series of elementary operations or motions that each of these men uses in doing the work that is being investigated, as well as the implements each man uses.
- *Third.* Study with a stopwatch the time required to make each of these elementary movements and then select the quickest way of doing each element of the work.
- *Fourth.* Eliminate all false movements, slow movements, and useless movements.
- *Fifth.* After doing away with all unnecessary movements, collect into one series the quickest and best movements as well as the best implements.

"This one new method, involving that series of motions that can be made quickest and best, is then substituted in place of the ten or fifteen inferior series that were formerly in use. This best method becomes standard, and remains standard, to be taught first to the teachers (or functional foremen) and by them to every workman in the establishment until it is superseded by a quicker and better series of movements. In this simple way, one element after another of the science is developed.

"In the same way each type of implement used in a trade is studied."

Taylor demonstrates that, in many industries and all types of work, improvements of an increase in productivity of three to four times, or even more, are common. But he is equally clear that there is a proper, principled way to introduce these changes to the workplace. After listing twelve different "details of the mechanism of management," he says that "Scientific management, in its essence, consists of a certain philosophy," which he summarizes as follows:

"It is no single element, but rather this whole combination, that constitutes scientific management, which may be summarized as:

- Science, not rule of thumb.
- Harmony, not discord.
- Cooperation, not individualism.
- Maximum output, in place of restricted output.
- The development of each man to his greatest efficiency and prosperity.

"The writer wishes to again state that: 'The time is fast going by for the great personal or individual achievement of any one man standing alone and without the help of those around him. And the time is coming when all great things will be done by that type of cooperation in which each man performs the function for which he is best suited, each man preserves his own individuality and is supreme in his particular function, and each man at the same time loses none of his originality and proper personal initiative, and yet is controlled by and must work harmoniously with many other men.'"

Although the term "quality management" had not yet been developed, Taylor's scientific management contains all of the goals and principles, and most of the approaches that have been elaborated and refined since that time. We can summarize Taylor's innovations in three key points:

- *Use observation, measurement, and experiments to improve work processes as well as engineering practices.* That is, we experiment to change the way we work, not just to define the technology and tools, then incidentally change the way we work.
- *Set standards from experiments, then manage the work to bring everyone to the level of the standard.* More experiments, including tests of suggestions from workers, can be used to improve the standards.
- *Management has a key responsibility to work with the workers, guiding with the knowledge provided by science, but doing it is a way that engenders cooperation.* In contrast, attempting to force improved methods on workers is a business disaster.

Taylor was the first to discover the principle that working smarter—applying scientific principles to improve the quality of tools and work processes—improves both productivity and quality.

Although Taylor and others succeeded in helping some companies transform along these lines, to the benefit of the company, the worker, and society, there were also failures. Taylor discusses these and identifies the principles that were disregarded in each case. It would be more than 30 years before any society successfully implemented Taylor's principles thoroughly, and it would happen half a world away, in Japan.

The Split After Taylor

Taylor's work was met with a mixed response. When companies applied his principles, and not just his methods, there was success. Companies that applied his methods without attention to his principles—trying to rush improvements and force them on the workers, for example—met resistance and strikes. The third point I identified in reading Taylor—that it is a management responsibility to guide workers to cooperate—was rarely well understood and even more rarely well implemented. Part of the source of this problem falls on Taylor. His writing makes it clear that he was an elitist, and that he thought educated managers had to lead uneducated workers. (He worked with many organizations that relied on large workforces of illiterate immigrants.) So, even though his own practice was cooperative and he saw engendering worker cooperation as an essential management responsibility, he was unable to convey this well to others.

The problem got so severe that Taylor's approaches were heavily criticized, and even banned in government offices. The primary complaint was that the methods of the time- and motion-study treated people like machines. A close reading of Taylor makes it clear that this would violate his principle. Nonetheless, the error was made, and Taylor was criticized for—and even identified with—the misapplication of his ideas.

A new field of endeavor, *human resources,* arose in the 1930s to deal with the mass strikes that occurred when Taylor's ideas were misapplied. At first, the field of human resources was generally highly critical of Taylor and scientific management. More recently, the field has been trying to apply our scientific understanding of people—the field of psychology—to identify what workers are best for what jobs, and how improvements can be made. This follows Taylor's recommendation that we focus on "the accurate study of the motives which influence men." ("Principles of Scientific Management," Chapter 2.)

Human resources has also proposed a model of the attitudes towards workers that goes a long way towards explaining the success or failure of the cooperative improvement of work processes that is essential to quality management. In 1960, Douglas MacGregor suggests that managers, whether we realize it or not, have one of two attitudes about workers, which he calls Theory X and Theory Y. Theory X holds that people do not like to work, that they have to be made to work, and that they need supervision to make sure that they work. If managers hold this view, it usually leads to conflict when management wants to make changes to work processes. Theory Y holds that people like to work and want to do good work. Management succeeds by removing barriers to good work, eliminating hassle, and rewarding good work and improvements. When managers at a company believe Theory Y, quality management initiatives and continuous improvement can succeed.

Shewhart's Scientific Management

Walter Shewhart was the next great figure in the history of quality management. After working as a professor of physics, he joined Western Electric in 1918, and then moved to Bell Labs at its founding in 1924 and remained there until 1956. The central focus of his work was *statistical quality control*—called at the time *statistical process control*. Now, it is generally referred to as *Quality Control*, though it is important to remember that quality control is simply inspection plus statistics. Shewhart also developed an approach called *Plan, Do, Check, Act (PDCA)* which is essential to quality management, both in statistical quality control and also more generally in continuous improvement. A third concept, the distinction between noise and data in a signal, was relevant to statistical quality control and also essential to the development of communications science.

PLAN, DO, CHECK, ACT (PDCA)

Shewhart created PDCA, a simple application of the scientific method that anyone can apply. In general, we can plan work that solves a problem, do that work, check to see if we got the results we wanted, and then take action to make use of what we learned. This can apply to trying to meet customer specifications and also to solving any other type of quality, effectiveness, or efficiency problem.

Because PDCA is so simple, it can be used by any engineer, and even by many mechanics and office workers. PDCA was promoted by W. Edwards Deming, and therefore is part of Total Quality Management and continuous improvement and its descendants, including Six Sigma and ISO 9000.

Many definitions and explanations of PDCA are available. Here is a very clear one, focused on the primary goal of quality management—delivery to customer specifications—from the Note in Clause 0.2 of ISO 9001:2000 explaining that the PDCA cycle applies to processes:

- *Plan.* Establish the objectives and processes necessary to deliver results in accordance with customer requirements and the organization's policies.
- *Do.* Implement the processes.
- *Check.* Monitor and measure processes and product against policies, objectives, and requirements for the product and report the results.
- *Act.* Take actions to continually improve process performance.

The beauty of PDCA is that we can apply it over and over again to the same subject, correcting our course to achieve better and better results. This type of repetition of a process is called *iteration.* We can apply PDA to change results—improving quality—and also to changing any type of process. Process improvement can do many things: increase quality, increase effectiveness, increase efficiency, and more. Because PDCA has been promoted so widely, it has many names including the Shewhart cycle, the Deming cycle, and the continuous improvement cycle. We will discuss PDCA more fully in Chapter 7: *The Quality Team* and see how various schools of quality management apply it in Part 3.

STATISTICAL QUALITY CONTROL

While PDCA can apply to both general efforts to improve quality and more technical engineering improvements, statistical quality control is the core of the more advanced engineering side of quality management. Shewhart developed the methods at Bell Labs, and they were included in best practices from the U.S. military for providing war materiel during World War II. Once again, the demand for large quantities of high-quality weapons drove the improvement in quality methods. In this case, the issue was the high cost and delays created by the requirement that every single item had to be inspected after manufacture and before delivery. Shewhart found a way to apply statistical sampling to the process of inspection. He showed how one could sample only a part of a production batch of a product, and yet be sure that the entire batch met customer specifications. He developed control charts and other tools for this work. Proper application of statistical theories about how to choose the sample, and then measurement of the sample to control limits that were tighter tolerances than customer requirements, could allow confidence that if, say, 10% of an entire batch was within the tighter tolerances, and the sample met certain other requirements, then the entire batch would meet customer requirements. Statistical Quality Control is fully explained in Chapters 6, 8, and 10.

Deming and Total Quality Management

W. Edwards Deming, a colleague and protégé of Shewhart, deserves a chapter of his own. He popularized and advanced Shewhart's work, and added significantly to it. In cooperation with Japanese scientists, engineers, and industrial leaders, he pioneered the development of Total Quality Management (TQM), which we discuss in Chapter 11. TQM was the first total solution to the quality problem that actually worked on a large scale. It includes PDCA, QC, and other quality methods which, used together, allow companies to sustain continuous improvement.

Quality in North America, 1920–1980

Our history has covered the main line of the development of quality management theory and practice from Smeaton to Taylor to Shewhart to Deming. As we close out the chapter, it is important to look at two other issues. How well or poorly were these ideas implemented, and why? What other ideas are related or relevant to the development of quality management?

PRODUCTIVITY, NOT QUALITY: NORTH AMERICA FROM 1920 TO 1980

Unfortunately, the United States has been very slow to adopt the quality management methods invented here. There have been three major barriers:

- *Economic success.* In times of high productivity and high profit, it can seem like there is no need for quality management. A factory can simply produce too much, and throw away what doesn't work. Or, if there is no competition, a factory can sell poor-quality products and then the company can make even more money in the repair business. In some cases *planned obsolescence*—the intentional creation of low quality to drive up billable repair costs—was actually preferred over the creation of quality products.
- *Elitism and management-labor conflicts.* In the United States, the efficiency improvements recommended by Taylor were taken as techniques, but his cooperative philosophy that could bring together management and workers was poorly understood and rarely implemented. Instead, the assembly line model, combined with elitism that valued workers less than educated managers, and a culture with lots of managers who operated by Theory X, led to management-labor conflicts and a corporate culture strongly resistant to change and process improvements.
- *A focus on quick profits.* Both Taylor and Deming emphasized that it takes two to five years to bring scientific quality practices to an entire organization. And it can be expensive. Although a quality management program often begins to pay for itself before completion, it doesn't necessarily improve profitability every quarter. The focus of quality management is long-term improvement, not the quick buck. Boards of directors and senior management are often pulled away from effective long-term profits by the need or the desire to show quick improvement in the bottom line. In addition, other techniques—such as mergers and acquisitions—can seem to show a better financial result more quickly, though they may not pay off in the long run if quality is reduced by the reorganization of the business.

Historically, Henry Ford focused on using the assembly line, elitism, and Theory Y management, and no one could challenge Henry Ford's methods because, as the saying goes, nothing succeeds like success. Although the demands of World War II led the United States to develop better quality management methods, U.S. economic success after the war left no incentive to keep U.S. industry on the path of continuous improvement. Fordism, as it was called, reigned into the 1970s. W. Edwards Deming and the engineers he taught had no way to convince U.S. manufacturing of the value of quality management. And, as we shall see, other countries were interested. It wasn't until the late 1970s that U.S. industry, facing stiff international competition, began to wake up and see the need for quality management.

OTHER DEVELOPMENTS IN SCIENTIFIC ENGINEERING AND MANAGEMENT

The notion of applying science to business had two other facets besides quality management: Research and Development(R&D), and Operations Research (OR).

The R&D Function

The notion of research and development as a way to generate new business began with Thomas Alva Edison. In 1876, he founded the Menlo Park research lab, an inventor's laboratory that focused on creating new products for sale and improving products for commercial production. Edison is most famous for the electric light and the phonograph. Most inventors create one or a few patentable ideas: Edison was able to create hundreds. R&D is performed by large corporations, by think tanks such as RAND (the Research And Development Corporation), and by special divisions of elite educational institutions. It is funded by both corporate sources and also by the government. Two key United States government agencies that fund fundamental research of significance for business are NASA, the National Aeronautics and Space administration; and the Defense Advanced Research Projects Agency (DARPA) which funds research leading to both military and civilian inventions, including the development of the Internet.

One of the most interesting R&D facilities is the Xerox Corporation's Palo Alto Research Center (PARC). Xerox was very aware of its origins in R&D— as a failing photography company, they had risked everything to bring an odd idea—dry photocopying—to market, and became a manufacturing giant beyond their wildest dreams when they succeeded. Chester Carlson, the inventor of xerography, had been turned away by 17 major companies, including IBM and 3M, before the Haloid Company picked up his idea, made it work, and became Xerox. Once it became a giant corporation, Xerox made an effort to stay connected to its innovative roots by founding PARC and funding basic research into new ideas. The earliest powerful personal computers and the computer mouse were both invented at PARC. But the Xerox management team wasn't able to see the value of its own R&D facility, and a young high school student—Steve Jobs—bought the idea and founded Apple Computer. If Xerox, with its large presence and vast resources, had used its own invention from PARC, it might have beaten IBM to the punch and become the primary developer of the personal computer. I mention this because, all throughout quality management, we see that good ideas can't go anywhere without the ongoing support of senior management.

Operations Research

One other field is the result of the application of scientific principles to business management, the field of Operations Research (OR). OR began in World War II when mathematical models were used to solve operational and logistical problems. OR includes a wide variety of business models and mathematical theories that can be applied to solve different problems. The common threads are usually systems models, advanced mathematics, the use of computers, and solutions that require both mathematical prediction and also intelligent human application and interpretation of the results of computer analysis. In this, OR parallels quality management. In both fields, the combination of human innovation and intelligence along with engineering or mathematical skill is essential.

Q-PRO

Science and Engineering: Which comes first?
Science and engineering have a very odd relationship with one another. In one sense, engineering is way ahead of science. Engineers invented and used domes and arches to keep bridges and buildings standing up centuries years before science could explain why domes and arches make buildings stronger. At the same time, new discoveries in science and the application of scientific methods move engineering ahead. Many people were trying to invent the airplane, but the Wright brothers succeeded because they first defined the scientific principles of aeronautics, answering the question: Why would a craft that was heavier than air fly? Then they built a plane to meet the specifications that their science required. Science and engineering—why things work, and how to make things work—keep pulling one another forward.

Conclusion: From Scientific Method to Quality Management

From 1750 to the present, people in business—especially in wartime—have sought to solve problems of effectiveness, efficiency, and quality. Theory has remained far ahead of practice. In fact, many of today's quality problems could be solved by proper application of Taylor's 1911 masterpiece on scientific management. We've seen the historical thread of the application of the scientific method to the solution of technical and business problems. The evolution of an art and science of management are separate from the art and science of engineering. In Chapter 3: *Defining Quality*, we will tie together all the ideas of the last several centuries into a clear, simple picture of what quality is and how we can bring it under management.

Q-Ball Quiz for Chapter 2

In Table 2-1, your answer sheet for the Q-Ball Quiz for Chapter 2, you will see a list of years or dates on the left, and two blank columns: one column for a **person, organization, or method** and one for an **idea.** Below the table, you will find two bulleted lists. The first contains 11 **people, organizations, or methods.** Place each item from this list next to the correct date in Table 2-1.

Table 2-1.   Your answer sheet for the Q-Ball Quiz for Chapter 2.

Fill in this table with your answers selected from the two lists below.		
Year	**Person, Organization, or Method**	**Idea (One line will have two ideas)**
1755		
1788		
Around 1850		
1876		
Around 1900		
1911		
1920s–1950s		
1920s–1980s		
1940s–1980s		
1947–present		
1970		

To finish the quiz, go down to the next list, the list of 12 **ideas.** Take each **idea** and place it next to the correct date, as well. Note that two ideas go next to just one of the dates. When you are done, you should have all of Table 2-1 filled in, with every person, organization, or method and all the ideas in date order.

This is the list of people, organizations, and methods to place into the middle column of the answer table.

- Assembly lines
- Frederick Winslow Taylor
- James Watt
- John Smeaton
- Operations Research
- The U.S. military
- Thomas Alva Edison
- W. Edwards Deming
- Walter Shewhart
- Walter Shewhart and W. Edwards Deming
- Xerox Palo Alto Research Center (PARC)

This is the list of ideas to place into the right-hand columns of the answer table above. (Note that two of these ideas both go into one of the answer spaces.)

- Applied the scientific method to engineering problems
- Defined the first industry standard
- Inspection, followed by rework or discard
- Plan, Do, Check, Act
- Scientific Management
- Statistical Quality Control
- The centrifugal governor
- The first feedback device
- The R&D laboratory for maintenance of innovative leadership
- The R&D laboratory for new inventions and viable commercial versions of products
- Total Quality Management
- Use of mathematics to improve business decisions

CHAPTER 3

Defining Quality

As engineers and managers, we are more intimately acquainted with these facts—and are therefore best fitted to lead in a movement—by educating not only the workmen but the whole of the country.
—Frederick Winslow Taylor, *Scientific Management*, 1911.

Now that we understand the history of the effort to create quality, both before the invention of quality management, and also in the last 250 years as quality management has developed, we can come up with a clear, practical definition of quality. This definition pulls together ideas about quality from four disciplines:

- *Philosophy,* which seeks to define quality.
- *Economics,* which looks at quality in terms of value and the fulfillment of needs.
- *Marketing,* which looks at customer value and the customer decision-making process. That is, how customers define quality and choose value.
- *Operations management,* which sees quality as conformance to specifications.

The definitions of quality from all of these different perspectives seem unrelated. But when we put them together in the right way, we see that they are related, and that uniting these perspectives gives us an effective approach to defining quality with the customer and delivering customer satisfaction in the creation of products and services.

Tying Together Many Ideas of Quality

As we discussed in detail in Chapter 1: *Quality Throughout History,* in each moment of life, each person interprets his or her experience at four levels: universal, cultural, social, and individual. That natural, automatic, largely unconscious evaluation leads to an inner sense that might be *I like this; this is good for me.* But, to generate that feeling, the experience must be satisfactory on all four levels and in all ways. Otherwise, the experience is soured. Also, our bodies and emotions are constantly changing. As a result, what we like for breakfast, we don't like for lunch.

For each person, the sense of value and quality is fleeting and everchanging. And no two people like exactly the same thing. It is no wonder that philosophers concluded that quality could not be defined. They were right: As people, we know quality when we see it, but we can't define it.

Yet no one can determine what is "quality" for me. That is, no one can tell me to like something, or tell me what to like. The philosopher's position is true, but it leaves the businessman stuck. For, to succeed in business, we have to deliver things to the customer that the customer values and perceives as valuable. So we have to define quality with the customer, and also attempt to do it for the customer, even if that is supposedly impossible.

There is a solution. In Chapter 2: *The Development of Quality Management,* we said that engineers know how to make things work even when scientists don't know why they work. The same is true for quality managers and quality engineers. We know how to define quality for and with the customers, even when philosophers and customers can't say why they like something or what they really mean by quality.

Economists and marketing folks look at quality from the customer's perspective. Economists see quality as that which gives value to customers by satisfying their wants or needs. Customers are willing to pay to receive that value, which sets up demand for products and services. That demand, in relation to available supply, determines market price. Marketing folks see quality as that which adds value, but then focus on the customer's perception of value. They ask if customers know what they want, and if they can be informed or persuaded to choose certain items. At that point, marketing goes in two directions. Some marketing campaigns focus on getting the message out so that customers can understand value and make a clear choice about their preferences. Others seek to influence customer preference, getting customers to change their minds about products.

Marketing and product definition involve a great deal of compromise because it is difficult to know what customers want. In figuring out what customers want—in defining quality—businesses face these problems.

- *The customer doesn't know, and no one can decide for the customer.* If the basic response to quality is, as the philosophers say, "I know it when I see it," then defining quality before the product is made and delivered is difficult. We solve this by including the *voice of the customer* in the quality definition process.
- *Each person has unique preferences, but we try to build one product that will satisfy many people.* This leads to choices about pricing and quality. For example, some people will buy something only somewhat satisfactory—even if it doesn't give them all of what they want—if the price is low enough. It also leads to product lines with options that make the product different for different customers, and to customization.
- *It takes time and money to work with the customer to define quality.* This can lead to a trade-off: higher quality items costing more.
- *It takes many types of expertise and analysis to work with the customer to define quality.* We may need to bring together everything from marketing experts to statisticians to computer programmers to figure out what the customer wants.
- *Customers react to experience, not just to products.* Our product or service is just one part of the experience. The customer may react to something else that is going on—not part of our product, and only perhaps under our control—in their experience of the product and of our company.
- *Customers are complicated.* Even one person is complicated. But, often, the term customer refers to a group of people. Mothers and fathers buy things for their babies and children. People buy gifts for one another. And all business sales have people in multiple roles. These include the purchase decision, use, and realization of benefits of a product. Even if we sell to a one-person company, that person buys and perceives value both as an individual and also in terms of his or her role in the company.
- *The real final value of the product is often not perceived before, or even soon after, the time of purchase.* If the product is something that the customer will use for a long time, then a customer's expectation that it will last five years can leave the customer dissatisfied when it only lasts three. Or, if a customer eats a big meal late one night and loves it, he may not be as happy when he gets on the scale in the morning.
- *Influencing customer desires through branding and marketing is an alternative to finding out what the customer really wants.* But after the customer has the product, a question remains as to whether we've really met a need, or only fulfilled a perceived need.

All of this complexity is very, very human. And most engineers are more focused on things and data than on people. That is not a criticism. Studies in the field of human resources show that everyone has a propensity for one of three areas—to work with people or animals; to work with data, knowledge, and information; or to work with physical objects and equipment. Each of these leads to a satisfying career. Psychotherapists, architects, and construction workers all add value to society. What makes quality management a challenge is that successfully defining quality and delivering it to the customer requires all three talents.

- We need people skills to elicit customer requirements and manage the team to deliver to those requirements.
- We need skills with data and information to form customer requirements into a technical specification and to plan the work.
- We need physical skills with things to actually make our products and services and deliver them.

The key turning point in this process is the specification of customer requirements. If we can take information that comes from outside—from the customer—and turn it into a clear definition of requirements that, if met, will deliver quality, then we've done what the philosopher cannot do—define quality so that we can deliver it.

This is not to say that the job is done once we have a specification of customer requirements in writing. Even if it is a good specification—clear, consistent, achievable, and having other qualities we will discuss in Chapters 4, 5, and 6— we are not nearly done. To finish, we have to do three more things.

First of all, we have to meet the operations management definition of quality: to develop and deliver the product or service in conformance to, or exceeding customer specifications and expectations. This is the arena where most quality management work has been focused. We need to:

- *Plan* how to deliver quality.
- *Do* the work following the plan.
- *Check* the work to make sure it conforms to specification, using inspection and quality control (QC).
- *Act* on the information we learn from checking, to eliminate error and deliver to specification.

We will use the PDCA cycle many times in delivering each product.

Secondly, we need to make sure that we are doing a good job—that we are testing everything and responding to what the tests tell us. This process is called *quality assurance (QA),* and it includes any and all activities—especially activities that cut across departments—that we do to make sure that we do our quality work right.

Thirdly, we need to deliver quality to the customer, and ensure that the customer receives not only the product, but the experience of quality, as well. Our goal can be—and should be—customer delight. Customer delight is what happens when the customer receives quality in the product and knows they got it, and is also delighted with every experience that they have with every representative of our company. Customer delight leads to customer loyalty, repeat business, referrals, and lasting success.

Pulling It All Together: The Practical Perspective

Figure 3-1 is a high-level view of the end-to-end process of quality management for development or improvement of a product or service. Checking, testing, and corrective action—the bulk of quality management work—are not included. The steps are:

- *Corporate planning* to define the basic parameters of the product or service, the target market, and the company's goals.
- *Requirements elicitation,* where we find out what the customer really wants and put it into a requirements specification.
- *Product, project, and quality planning,* when we define what we are making, how we will make it, and how we will make sure that we do a good job and deliver a result that conforms to specification. This includes identifying other requirements, such as the company's own requirements, the requirements of other stakeholders, and industry or other standards.
- *Developing the product, with QC and QA to ensure success.*
- *Managing customer expectations and change requests* so that the customers will be happy when they get what they asked for.
- *Delivering the product with quality,* where we deliver the product so that the customer realizes value from its quality, and we follow through to close the sales cycle and continue or close the relationship with the customer, ensuring the quality of our reputation with the customer.

The figure shows that the process moves on two tracks, an upper track of customer communications, and a lower track of product development. Delivering quality requires continuing work on both tracks.

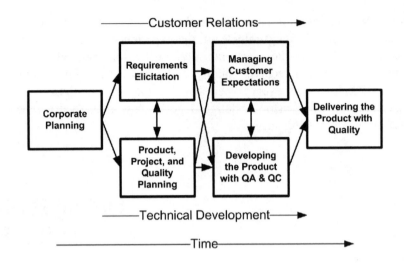

Fig. 3-1.   A high-level view of the end-to-end process of quality management.

REQUIREMENTS ELICITATION

Requirements elicitation is a topic worthy of a full book of its own. In summary, it is the practical solution to the paradox—we need to know the customer's definition of quality, and the customer experiences quality, but can't define it. We will discuss the requirements elicitation process in a bit more depth in Chapter 6: *Defining, Planning for, Controlling, Assuring, and Delivering Quality.* For now, we can make these key points:

- Requirements elicitation can be thought of as defining the specific qualities, of a specific product, service, or solution, that add value.
- Requirements elicitation is a dialog. It needs to be structured by the provider, and draw forth new understanding from the customer.
- The core of the dialog is to ask and answer the question, "What makes it good?" We ask everyone involved in every way we can, including as much detail as is feasible.
- The dialog must recognize the constraints of the effort. If we give the customer a free hand without regard to price or time, they will end up asking for something that is not affordable or feasible.
- The dialog must have a beginning, middle, and end. The beginning may be a question, a problem, or a proposed product or service. The middle is an

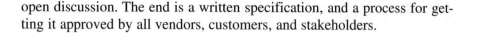
open discussion. The end is a written specification, and a process for getting it approved by all vendors, customers, and stakeholders.

We must take the initiative and responsibility in creating and completing the requirements elicitation dialog. Good requirements elicitation is essential to defining and delivering quality, and it is irreplaceable. If we don't do it well, all we have is excuses, instead of quality and success. Excuses such as, "I didn't know the customer wanted that," or "We thought we knew what the customer wanted," or "The customer didn't tell us he needed it done this way, or by this date" are the result of poor requirements elicitation. In these situations, we either fail to deliver anything, or we are stuck with expensive rework. If we deliver value at all, we deliver a much lower ratio of value to cost. When we do good requirements elicitation, we get a good requirements specification. That allows us to take control of the development of the product or service, using methods from operations management and quality management to deliver a product or service that conforms to the specification.

TYPES OF SPECIFICATIONS

There are many different acceptable final results of the dialog of requirements elicitation:

- *A solution.* The customer may want a solution to a problem, and it is up to us to define what will solve the problem and how we will make it.
- *A capability.* The customer may say, "the product or service needs to make me able to do X," and then leave it to us to define what will do that, and how.
- *Functionality or feature.* The customer may say "the product or service needs to do (or have) this, and this, and this."
- *The product or service.* The requirements elicitation may end with a full specification of all interfaces between the product and the customer, defining the product or service as the customer sees it. The team will still need to define the technical components of the product.

In requirements elicitation, we must either set one of these goals and steer towards it, or educate the customer and make a collaborative choice where we agree on the type of specification we are going for. Such a decision might be, "The customer states that, as long as the product is capable of X, they don't care how we do it." Or, "the customer and the vendor agree to work together to fully define all features of the product that the customer will see and interact with, and then the vendor will develop the internals to meet those requirements."

CONFORMANCE TO SPECIFICATIONS

Once we have a customer's requirements specified, we move into the realm of quality management defined by the field of operations management: delivering a product that conforms to specifications. In the past, some companies tried to use this as the only definition of quality, and that created a huge gap between quality and value. The gap is created when the specification doesn't really represent what the customer wants and needs. More recent definitions of quality management generally define the goal of quality management as delivering to specification, and also meeting or exceeding customer wants, needs, and expectations.

Conformance to Precise and Imprecise Specifications

If a specification is good—clear, consistent, unambiguous, prioritized, and in other ways usable—then conformance to a specification is testable. To the degree that there are errors or ambiguities in the specification, testing to see if the product conforms is difficult. Some types of specifications are precisely defined and measurable. Whenever it is possible to make the specification precise and measurable we should do so. That way, all parties can objectively agree on whether the specification was met. However, some specifications are inherently imprecise. If our business is to run children's birthday parties, then the ultimate requirement is that the kids and their parents liked the party. We can specify a lot of that requirement, but we could never specify all of what is required. Conformance to imprecise specifications requires good customer relations—effective two-way communications with the customer that includes feedback so that we can correct our course and deliver success. That way, we can meet requirements—satisfy wants, needs, and expectations—even when we can't fully specify them.

Controlling Changes to the Specification

What if the specification changes during the course of the project or during the life of the product or service? Requested or essential changes to specifications are a common occurrence, and our quality management system must deal with them. We do this through a method called *change control*. In change control, we define requested changes to the specification, evaluate them, approve or reject them, and communicate our decision to the customer. If we accept the change, we must update the specification and communicate the change and all of its consequences—time, cost, risk, testing, and other factors—to the team and all stakeholders.

Customer Specifications, Stakeholder Specifications, and Standards

We need to realize that a customer specification is not a complete specification. The specification must define what will satisfy the customer, but there are three other things it must do as well:

- *Meet requirements of all stakeholders.* Anyone who is affected by a project, product, or service is a stakeholder. As we define a product, we may find that, in addition to the primary customer, others, such as service and repair groups, shippers, and vendors, have requirements for the product as well.
- *Meet standards and regulations.* Organizations that check or enforce standards and regulations—through audit or other mechanisms—are a special category of stakeholder. Part of our work as quality managers is to identify any standards and regulations that the product, service, or creative process is required to conform to, or are advantageous to us, and include them in the specification.
- *Define all external interfaces.* A systems model of a product or service will identify every interface—every surface of the product that has contact or communication with any other object or system—and define what does and does not cross that interface. Whenever an interface is left undefined or poorly specified, value is lost and cost increases. Proper communication and transmission across interfaces assures desired results. If you've ever seen garbage come out your printer, or seen scrambled text in an email, or found that you ordered one item and got another, a systems person could say either "the interface was poorly defined," or "communication or transmission failed at the interface." If it happens repeatedly, it's a matter of poor definition.

A specification that meets these requirements is a complete product (or service) specification. If we specify sub-components of the product or service, we are doing the same for each sub-component, and therefore we are also defining internal interfaces—interfaces between components of the product.

Achieving Quality: Managing Error

To complete our definition of quality, we need to turn quality on its head. The opposite of beauty is ugliness. The opposite of good engineering is disorder.

The opposite of quality is error. As a result, quality management is error management. It is very easy to see that if we define error, bring it under management, and reduce or eliminate it, we manage and deliver quality at the same time.

In this initial discussion, I'm not distinguishing error from defects. I'm treating the two as more or less the same thing for now, and I'll use whichever term is more appropriate. You'll learn the precise difference when we discuss statistics in Chapter 10.

Table 3-1 shows the most significant consequences of error in each of the major processes of quality management defined earlier in Figure 3-1.

Table 3-1. Consequences of error in the quality management process.

Consequences of Error in the Quality Management Process	
Quality Management Process	**Significant Consequence of Error**
Corporate planning	An error in corporate planning can cause us to create a product or service that won't work at all, or won't be optimal for the company.
Requirements elicitation	An error in requirements elicitation leads to an incorrect specification. If we don't catch the error, we deliver to specification but don't satisfy the customer. If we catch the error, it will cost us ten times as much to fix it later as it would have cost to fix it during the elicitation process.
Product, project, and quality planning	Each error here costs ten times as much to fix during the course of the project as it does to prevent the error by proper planning. Major errors may create a situation where we cannot deliver a quality product on schedule affordably.
Developing the product with QA and QC	Each error here is one defect in one instance, some instances, or all instances of the delivered product.
Managing customer expectations and change requests	An error in managing change requests can lead to complete project or product failure. An error in managing customer expectations can lead to delivering a product that meets specifications but does not satisfy the customer. This can lead to a major loss for the business, where all the effort of product development may be spent with no value realized.
Delivering the product with quality	An error here can lead to failure at the end of a project or loss of cusomer satisfaction and repeat business.

We bring error under management by defining it. We define error by defining a complete product (or service) specification with tolerances. Anything that falls outside of tolerances is error. Of course, it is essential that our specification be error-free and remain free of error. So, we need quality assurance to ensure our methods of defining the specification lead to a high-quality specification. After that, we need change management and more QA to make sure that the specification remains error free, and that the customer's expectations are all defined in the specification. If we do all of that, then we can meet customer expectations by developing a product that conforms to specifications—a product that is error-free in relation to its written specification.

Many approaches to quality management limit the notion of error to processes that relate to meeting a specification or standard, and defects to failures of the product to meet its specification. The broader approach in *Quality Management Demystified* is derived from the sense of the word "total" in Total Quality Management (TQM). Every process, document, and interim or final deliverable is subject to improvement through PDCA. A good requirements specification reflects not just the customer's definition of value. We can increase the quality of what we offer and add value for our customers by improving requirements elicitation and customer communications. When the requirements specification is as good as it can be, then what is usually considered the core of quality management—conforming to specification through checking and process improvement—comes to the fore. If the product meets the specification, we can do more to ensure the customer's receipt of quality and perception of value by following through with quality delivery for customer delight.

This total approach to quality—using quality management to reduce or eliminate all errors in all processes that affect quality and/or value—requires both a human effort and a technical effort.

THE HUMAN SIDE

Reducing error is both an individual effort and a team effort. We want each person on the team to care about doing good work and delivering error-free results. We also want to make sure that they have the necessary skills and are empowered to take actions to see, define, speak about, and eliminate errors. We need to create a culture that asks each team member to boldly speak the truth about eliminating error and about naming and fixing problems promptly when we see them. And, when people are working alone, we want them to have the confidence to do whatever it takes to eliminate error and deliver work that meets or exceeds specification.

Even with the best individual workers in the world, quality must still be a team effort. It is always harder to see one's own errors than to see someone else's. Some errors are the result of a misunderstanding or misperception. If I misunderstand the specification, then, when I check my own work, the same misunderstanding will lead me to think that there is no problem.

EYE ON THE BALL

Specify "Next Tuesday"

Whenever I teach a class, I use this example to illustrate assumptions about specifications. Say I'm teaching on a Thursday. I ask the class, "How many days from now is next Tuesday? Raise your hand if you think next Tuesday is five days from now?" About half the class raises their hands. "How many of you think next Tuesday is twelve days from now?" The other half of the class raises their hands.

This is an example of an error a person could not correct himself. If a person thinks he understands a term in the specification such as "next Tuesday" correctly, he won't question if it might be wrong, and he won't catch the error.

The lesson: We can't catch our own assumptions, so we can't catch all our own errors.

As a result, we need ways of working as a team to ensure quality. Working together to deliver quality, we can set up independent checking, so that an error created by someone with an assumption or a certain perspective is caught by someone who doesn't share the same way of looking at things. We call this *independence* in checking and testing.

There is a fundamental challenge here. A person's best effort to do good work requires a strong personality, what some would call a strong ego. Letting others check and correct that work, and then being able to fix errors that others find, requires humility, an absence of ego. The fundamental challenge of building a good team of good people requires strong personalities, and also the ability to leave one's ego at the door when entering a meeting for quality review, inspection, or testing.

There is an easy way to do this. We can tell our teams: No matter how good each person is, we all make mistakes sometimes. Error is part of being human. Each of us will do our best to make as few errors as possible, but some error is inevitable on every job. At that point, the question is: Who do you want to find the error? Would you rather have the team find it, so you can fix it? Or would you rather have the customer find it?

THE TECHNICAL SIDE

The technical side of error management includes both design to reduce error and also checking, a general term for all types of inspection, review, and testing to find errors. We examine or test attributes of the product or process and compare them to the specification. When the measured result falls outside the range of tolerance for specified value of the attribute, there is an error. In inspection, we examine every instance of the product. In Statistical Quality Control (QC), we examine a sample and then apply statistics to determine how much error is probably present in the whole population. This checking is part of the PDCA cycle, and we take appropriate action based on the results of the checking.

Test design, measurement techniques, sampling, and analysis of test results can be extremely technical. Taylor argued that workers could not do it—that it had to be done by specialists. Forty years later, Shewhart showed that workers could be trained to do it if they learned PDCA. Starting in the 1950s, Japanese industry adopted PDCA with Deming's help, and we saw TQM—the first long-term quality management efforts—across an entire industry. Since that time, continuous improvement has been possible, but its application has been hit or miss. Some companies maintain systems that steadily reduce error, year after year. Others treat quality management as a fad, to be replaced by some other initiative that improves the quarterly bottom line, and cycle in and out of delivering quality in perpetual oscillation.

Our Case Study:
The Ham-and-Cheese Sandwich Defined

A good ham-and-cheese sandwich:

- *Is a solution* to the problems of hunger and lack of enjoyment.
- *Gives the customer the capability* to take it with him or her, solving the problem of the need for individuals to eat later in the day at a remote location.
- *Shows functionality* when served at business meetings, because people can eat it with one hand while taking notes with the other.
- *Can be defined as a product* consisting of two slices of bread, with appropriate amounts of ham and cheese between them, and possibly other foods or condiments as well.

A ham-and-cheese sandwich is a flexible product. It can be pre-packaged with uniform contents, prepared on a platter with a number of variations, or custom prepared, allowing each person to design his or her own and have it made to order. It can even be delivered in component parts for the customer to assemble. Common variations include:

- The type of bread.
- The type of ham.
- The type of cheese.
- The presence, absence, and quantity of various additions and condiments, including mustard, mayonnaise, salad oil, lettuce, tomato, pickles, red or white onions, and jalapeno peppers.

Cultural and stylistic variations are common. There is a California-style wrap, where the bread is replaced by a rolled tortilla, and the high-end product, the French *croque monsieur*. There are variations to meet customer needs, such as a vegetarian ham-and-cheese sandwich made with tofu instead of meat, and a sandwich with substitute soy cheese for people who don't eat cheese because they have lactose intolerance.

Measurable components of quality for a ham-and-cheese sandwich include:

- The amount—measured by weight—of ham and cheese.
- Its nutritional composition.
- A label with ingredients, required by the FDA.
- Freshness, which, when it degrades, is at first an issue of taste, and later a health issue.
- A process measurement of the temperature at which all ingredients and the final sandwich are stored.

As an exercise, design a ham-and-cheese sandwich suitable for each of these occasions.

- A lunch at a seminar, where 100 people need to pick up their lunches quickly, but people want options of type of bread, ham only, no cheese, and so forth.
- A fancy party for a few friends.
- A picnic.
- As an item sold at an all-night convenience store.
- As an item sold at a deli counter.
- As an item on the menu of an upscale restaurant.

- As a gift-box of supplies, ordered from a catalog or website, and shipped to a customer for assembly.
- At your own birthday party.

In each case, maximize quality at reasonable cost.

Conclusion: Making Quality Real

In Chapter 3: *Defining Quality,* you have seen the entire process of quality management laid out from end to end. First, we defined what product or service will add value for the business. Then we used *requirements elicitation* to solve the problem of finding out what customers want, even though they can't usually put it into words by themselves. Most quality problems are a result of errors in that step, leading to a poor requirements specification. If we have a good requirements specification, then we can plan how to create something to meet the specification, do the work, check it with inspection and QC, and act to correct errors, and prevent repetition of errors. The PDCA cycle applies for the entire product, and also at a detailed level as we define each process and correct each detail. We can view the work of conformance to specifications equally as an effort to achieve quality and also an effort to eliminate error.

Some companies have pursued and achieved quality consistently for many years. The trick is to set up systems that allow in very little error, and then eliminate the error that remains. The purpose of this is to satisfy the customers' wants and needs. As we shall see in Chapter 4: *Quality for the Customer,* building a reputation for quality is excellent for your customers and your business.

Q-Ball Quiz for Chapter 3

1. Which of these fields defines quality as "conformance to specifications"?
    (a) marketing
    (b) economics
    (c) operations management
    (d) philosophy

2. Which of these is a core question asked during requirements elicitation?
   (a) What makes it good?
   (b) What does the customer want?
   (c) What fulfills customer needs and expectations?
   (d) How do we eliminate error?

3. Quality management is synonymous with
   (a) quality assurance.
   (b) requirements elicitation.
   (c) inspection.
   (d) error management.

4. Which of the following is *not* a way of managing the fact that we make many identical products, but customers all have different preferences?
   (a) Customizing the product.
   (b) Standardizing the product, and accepting that we won't please everyone.
   (c) Requirements elicitation.
   (d) Making multiple versions of the product with different features.

5. Which of the following is *not* a result of a poor quality requirements specification?
   (a) We meet the specification, but don't satisfy the customer.
   (b) Costly rework.
   (c) A measured result falls outside accepted tolerances, indicating presence of an error.
   (d) More problems in production.

CHAPTER 4

Quality for the Customer

This means increase in prosperity and diminution in poverty, not only for their men but for the whole community immediately around them.
—Frederick Winslow Taylor, *Scientific Management*, 1911.

In business, quality means little if it is not quality for the customer. In fact, the voice of the customer should be essential in the definition of quality. Yet business has had a hard time integrating the customer into the quality process, or, when we do, we sometimes forget the needs of our own business. In this chapter, you will see how to bring the customer in, and take care of both the customer and the business through quality management.

EYE ON THE BALL

Quality Must Be for Customer and Company
Sometimes, quality management improvement efforts lose sight of the goals that matter for business. We can prevent this by keeping one eye on the customer, and the other on the bottom line. Each quality improvement should do one or both of these: Deliver more value to the customer or improve the bottom line.

Usually, when we deliver more value to the customer at the same cost, we do improve the bottom line. Greater customer value makes it easy to increase market share by winning new customers and reducing sales costs by getting more repeat business and referrals. In addition to increasing revenue by adding quality, quality management efforts can also decrease cost by improving the quality of processes. Process improvement leads to greater effectiveness and greater efficiency.

Quality for the Customer

What do we mean by quality for the customer? We need to answer this question from three perspectives:

- *The customer perspective.* Quality for the customer means that, in selecting and buying the product or service, the customer has a hassle-free experience, and in using the product or service it meets or exceeds expectations for as long as they want it to. If we are providing quality for customers, then, at any moment during or after the process, they would buy more from us or recommend us to others.
- *Customer quality from the business perspective.* Key issues include: Identifying the target market and the needs of that market, establishing effective communication with customers or customer representatives to develop a good requirements specification, providing high-quality sales and customer service so that the customer likes the company and the interaction, as well as the product or service, and doing all of this affordably.
- *Customer quality from the technical perspective.* Technical groups can only deliver quality to the customer if the requirements specification is a true, complete, clear representation of the wants, needs, and expectations of the requirements of all targeted customers and all stakeholders.

In working with the customer to define quality, we need to work very differently when looking at consumers or business customers.

CONSUMERS

Let's first look at defining quality for consumers, and then at delivering quality to them.

What is Quality in Customer Service?
It's nice to talk about defect-free products but, realistically, mistakes will happen. Defective products will slip through, shipping and billing errors will happen, and customers will forget to pay their bills and need to be notified. The question is: How do we do these things well? Really good customer service responds to a mistake with a full apology, corrective action, and sometimes an additional gift. It is so unusual that, sometimes, making and correcting a mistake wins us a loyal customer.

How do you handle customer service problems now?

How could you do it better?

In defining quality for consumers, we need to address these issues:

- *Identifying customer groups.* We identify our market by segmenting it into customer groups. Usually, we define them by age, gender, and where they live. But we can use other key factors as needed.
- *Describing each customer group.* We describe each customer group by identifying: the *purchaser,* who decides to buy and pays for the item; the *user,* who actually uses it; and any other stakeholders who may be involved in the purchase decision or who need to be satisfied with the product or service. We then identify the customer's key need or problem to be solved and other elements of their interest in the product or service.
- *Defining the customer requirements specification.* At this point, we are ready to work out the details of what the customer really wants, and to describe it in such a way that our technical team can create the product or service and our marketing group can figure out how to promote and advertise it. We do this by working with representative customers or sometimes, by working with *customer representatives.* For example, a savvy marketing department might be able to define a product modification or an initial prototype of a new product, so that we save money by not working directly with the customer. The risk of using customer representatives is that, if they are wrong, we make a product that we think the customer wants, but which is not really valuable to the customer.

Defining consumer needs involves working with the mystery of personal satisfaction. Fads, styles, and the ability to influence choice—called creating a market—through advertising and establishing brand loyalty are all part of the process. Much of this operates outside of, or even against, logic. For example, an item might not be selling well until we raise the price. How can we increase

MIS-Q

Increase Price, Increase Sales

There was a European beer being imported to the U.S. by a small brewery. They kept trying to get a foothold in the market. When they couldn't, they lowered their price—again and again. Finally, running out of money, they sold out to a major U.S. beer manufacturer and distributor.

This company took a different approach. They didn't change the beer. They just raised the price, and the beer sold well as a "premium import beer."

The lesson: Price affects the customer's perception of value. Sometimes we can sell more just by increasing our price, because it places us in the right market niche.

sales by increasing the price? Well, if the previous price was so low that people thought the product was cheap, they wouldn't buy it. Raising the price can reclassify the product in the customer's mind as high end, exotic, or deluxe, and actually increase sales.

Once we define customer requirements, it is up to internal work—often technical or engineering—to meet that specification and prepare a product the customer will like. Then we must return to the customer and deliver the product effectively. This is the work of marketing, sales, and customer service with a focus on customer delight. Customer delight includes:

- *First and foremost, an ability to genuinely see things from the customer's perspective.* Many companies come up with ideas and gimmicks that seem to favor the customer, but really focus on the company's bottom line. The customer can always tell.
- *Appropriate marketing and advertising* that promotes the product honestly, features its benefits, and gives customers a good basis for deciding to buy.
- *A sales process, whether automated or personal* that is hassle-free and appropriate.
- *Customer service, technical or product support, and continuing customer relations* that are hassle-free, give customers what they want and need, and build a lasting relationship.

Customer delight, as a corporate goal, means that customers enjoy and appreciate every aspect of the product and every contact with the company. The business value is customer retention, repeat business, referrals, and testimonials. Customer delight boosts both sides of the profit equation. Increased sales increase revenue while repeat business and referrals reduce the cost of each sale.

BUSINESS CUSTOMERS

Defining the needs of business customers and selling to them is a bit more straightforward than it is when we sell to consumers. Value to a business is that which improves the bottom line. Our first goal is to answer these questions:

- How does our product or service improve the customer's bottom line?
- What roles or job titles define the decision makers in the selection of this product or service?
- What are the key factors in the decision? If there is direct competition, what would make us better? If there is not, how can we demonstrate to our customers the value of what we do?

From this point, marketing can define the target market and decide how to reach the right people while the process of gathering customer requirements gives us the information we need to develop or improve our product.

The Customer/Quality Divorce

So far in this chapter, we've talked about how to do quality work in customer relations. In fact, it's pretty simple and obvious, once you think about it. Now, we need to talk about how it has happened that many companies—including many companies with extensive quality programs—miss the mark, by an inch or a mile, and improve quality without adding any real value for customers or for the company.

Quality on Trial illustrates this with two examples on pages 13–16:

- A computer chip manufacturer spent $150 million to create a chip that was defect-free and would last 15 years. Unfortunately, they didn't consider that the customer knew that chips would be outdated in 2–3 years. The customer went with another vendor who had focused on a chip that lasted 2–3 years and cost less.
- A local transit authority in Britain is quoted as saying, "It is impossible for them [bus drivers] to keep their timetables if they have to stop for passengers." Keep the image in mind of empty busses sailing past waiting passengers the next time you are looking to see if your measure of quality really serves the customer.

Such stories seem amazing, but they happen quite often, and for just a few basic reasons related to corporate culture and corporate quality programs. The common root causes of these problems are:

- *Internal managers and workers have no connection with the customer.* Inside a large company, workers receive praise for, and pressure to, deliver results defined by other managers. Sometimes, these goals are part of what is needed to satisfy customers. But all too often they are either internal and meaningless to the bottom line, or incorrectly defined, so that they actually reduce customer satisfaction.
- *Companies don't promote an empowered culture with customer focus.* Studies show that many workers believe that their own companies don't care about the customer or don't provide quality to the customer. When a worker believes that, the worker probably has solutions to customer quality problems, but is convinced that the company won't listen.
- *Companies don't use enough common sense and process mapping.* In addition to workers making the common-sense connection, tools such as process mapping and system modeling can show which processes are really related to customer and company value. Companies could use these and then focus on improving key processes, instead of telling managers and workers to define all processes in a department.
- *Quality improvement begins by reducing cost, but ultimately offers more by improving value.* The initial focus of a quality improvement program is to reduce cost by eliminating unnecessary processes and reducing wasted time and materials spent creating defective products. Many companies achieved those goals, but didn't know where to go next. And they couldn't see where to go, usually because the company failed to communicate a vision that linked every employee and every process to both the delighted customer and the improving bottom line.

For the above reasons, quality improvement programs can lose focus or become misdirected and, as a result, disappoint customers and damage the bottom line. I've seen two other ways in which a quality program can actually succeed at delivering better value to customers, but still fail to improve the bottom line:

- *We're doing a great job, but nobody knows it.* I've seen some companies that are doing quality work for their customers, but do not have effective marketing. As a result, there are plenty of customers out there who would like what they do, but simply don't know that they are there. The business can fail due to insufficient marketing even when the company delights the customers it has.

- *We improved, but so did everyone else.* Many executive projections fail to ask, "What if the competitors do a quality improvement program just as good as ours?" Say three companies share nearly all of a market about equally. One company evaluates a quality improvement program. The team—or more likely, a consultant—says, "This quality program can reduce costs due to waste by 10% in the first year. Our competitor's products are similar to ours. We can cut price by 5% and increase both our market share and profit margin at the same time." He then shows a rosy picture of what it will look like to have 50% of the market and a larger profit margin. What no one asks is, "What do we get if our competitors do an identical quality improvement program?" The answer would be diddly squat. If all three companies improve equally, then, a year from now, there will be a price war. Customers will be better off, getting the same quality at lower price, but the businesses will have the same customer market share and the same profit margin.

Historically, Japanese companies got a head start in steadily delivering quality. We'll discuss this in detail in Chapter 11: *Total Quality Management.* Japanese productivity and quality climbed steadily starting in the 1950s and, by the end of the 1970s, North American businesses were in big trouble. This led to over a decade of major quality improvements. However, by the early 1990s (for the reasons we've discussed above) the first companies were seeing that quality improvement did not always mean more productivity and profit. The issue of quality for the customer really came to the fore in the early 1990s, and terms such as customer delight and the voice of the customer began to shift the American business focus from internal quality improvements to total quality for the business and the customer.

Unfortunately, by the mid-1990s, several other things got in the way. Executives who had been promised higher market share and greater profit margins were disappointed. If we follow the example from the last bullet above, we can see how this happened. What the executives didn't see is that, in doing a quality program, they had done the right thing for the wrong reasons. They thought they would get ahead, and were disappointed. What they didn't see is that maintaining quality improvement was simply essential for survival. Companies that did not launch improvement programs never saw the increased efficiency and went out of business. They went the way of Zenith televisions and U.S. Steel.

So the executives survived, but they were dissatisfied. In fact, they were dissatisfied customers of the quality movement, and they were feeling a lot of pressure to fix the bottom line. This led to a flurry of profit by merger and acquisition, and a focus on branding to gain customer loyalty through advertising. Process re-engineering often focused purely on cost, sometimes at the expense of quality, and often at the expense of employee loyalty. Some industries faced heavy costs from the need to prevent disaster due to lack of computer and equipment compatibility after the year 2000—the Y2K bug.

And then came the Internet, a whole new way of business, an apparently ever-expanding bubble of opportunity, and then the burst of that bubble. That brings us to the year this book is written—2005—with American industry first losing its focus on quality, then interfering with quality through mergers, acquisitions, and downsizing and, all too often, looking for the quick fix and the quick buck due to a focus on the quarterly profits announced to stockholders. Now, the Internet is doing for soft skills—software development and customer service—what the Japanese success with TQM did in manufacturing in the 1980s. Business is moving overseas. And a new vision for quality improvement could be the solution now as it was in 1980, as we will see in Chapter 19: *Global Quality in the 21st Century*

The Voice of the Customer

The voice of the customer is an approach that developed in the early 1990s in North America among both the promoters and critics of Total Quality Management. It has since been adopted by the more recent Six Sigma quality movement. "The voice of the customer" is a simple catch-phrase to remind us that we need to be certain we are addressing customer requirements—not purely internal requirements—or our idea of what the customer wants. This issue can be raised in product design, and also in troubleshooting quality problems. It allows us to ask:

- *Do we really know what the customer wants?* Or do we need to find a way to gather or check customer requirements?
- *Do we know if this issue really matters to the customer?* Let's make sure before we fix what isn't broken.
- *Do we really know the customer's view and issues on the problem we're working on?* Or, are we doing the customers' thinking for them, instead of listening to them?

As we discussed in Chapter 3: *Defining Quality,* the customer is not the only group that has requirements. We need to define and meet the requirements of all stakeholders, including applicable regulations and standards.

Conclusion: Quality for Customer and Company

For the most part, delighting the customer by meeting all needs, wants, and expectations in our product and providing a high-quality customer service experience is beneficial for the company. But, as always, success requires balance. It is possible to become too customer focused. All of our quality efforts should focus on delighting the customer and we should be able to demonstrate how each quality improvement will benefit the customer, or not reduce customer value while increasing productivity and profitability.

There is one re-engineering model that is sometimes too focused on the customer. The concept is that we can replace management requirements with customer requirements. The slogan says that we eliminate management and give the customers what they want. This works in the limited circumstance where our business customers provide us with complete specifications. However, in all other cases, we have management work to do because, as we discussed in earlier chapters, the customer can't define quality by himself. We must assist in the process of quality definition, then ensure its delivery. That is our essential function as quality managers.

At the same time, executives and quality managers also need to pay attention to the needs of our own business. The cost of a quality program has to be affordable and reasonable. Quality programs must not interfere with or take away from routine business operations and other important initiatives. And we can't let the customer's eyes grow bigger than their pocketbook when we ask them what they want, or they will define requirements that cannot possibly be delivered at reasonable cost—even though we've led them to expect such miracles.

We have finished our journey through the history of quality and the major issues facing quality management. Now, we turn to Part 2: *Quality Essentials,* where we define the crucial ideas and methods of quality management.

Q-Ball Quiz for Chapter 4

1. In categorizing consumers into customer groups, which of the following is *not* important?
   (a) Identifying the buyer/decision-maker.
   (b) Identifying the customer/user.
   (c) Contacting every customer.
   (d) Defining the primary need of the customer.

2. The most important issue in customer delight is
   (a) being able to truly see things from the customer's perspective.
   (b) delivering a product that totally satisfies the customer.
   (c) ensuring that the customer likes our company.
   (d) guaranteeing increases in productivity.

3. Which of these is *not* a reason that quality improvement programs become separated from customer value?
   (a) Internal managers and workers have no connection with the customer.
   (b) Big companies just don't care about their customers.
   (c) Companies don't promote an empowered culture with customer focus.
   (d) Companies don't use enough common sense and process mapping.

4. Which of these was a source of disappointment in the results of quality improvement programs for North American executives in the early 1990s?
   (a) Promised increase in market share did not happen because competitors improved quality, too.
   (b) Total Quality Management could not be adapted successfully to work in the U.S.
   (c) The dot-com bubble burst.
   (d) Customers didn't care about getting better products.

5. "The voice of the customer" can best be described as
   (a) a method that was defined by the developers of Six Sigma quality management.
   (b) a catch phrase that reminds us to be sure we know what really matter to the customer.
   (c) a structured tool for gathering customer requirements.
   (d) a way of solving quality problems.

PART TWO

Quality Essentials

Now that we have seen the history of the problems and solutions of quality management, it is time to take a look at our toolkit. What activities can businesses engage in to bring quality under management? To bring something under management means to define it, plan what we want to do about it, do it, then track and control what happens, trying to keep things going according to our plan. In quality management, defining and planning are often the hardest parts. In cost management, we don't need to define money—the government does that by giving us dollars, yen, and euros as units of measure. In time management, we have hours and seconds as a defined unit of measure. In quality management, we have to specify what we will measure, and there are often many different things we want to measure. Once we've defined what quality is, then anything outside that range is error. At that point, we do a lot of quality management by bringing error under management and reducing it to acceptable levels.

We will begin with Chapter 5: *Key Quality Concepts,* where we define requirements, specifications, standards, and error; and then look at processes of checking—review, inspection, and testing—the basic processes of finding error. We will also look at how to prevent and remove errors, and then look at the consequences of leaving errors in the product, which leads us to the topic of the cost of quality. We close Chapter 5 with a look at why it is hard to keep a focus on quality in the business environment.

Once the key concepts of quality are covered, it is easy to understand the core quality methods. In Chapter 6: *Defining, Planning for, Controlling, Assuring, and Delivering Quality* we look at each of these five methods. Most books begin with Quality Control (QC) and Quality Assurance (QA), but, in Quality Management Demystified we want to prevent common confusion about them, so we've laid a solid groundwork before introducing them. By the end of Chapter 6, you will understand QC and QA in the larger context of bringing quality under management and delivering high-quality products and services to your customers.

In Chapters 7, *The Quality Team* and 8, *Quality Engineering,* we introduce the two essential sides of managing quality—the human side, through a combination of teamwork and individual effort, and the technical side, through measurement, comparison, the scientific method, engineering, and statistics. Chapter 9: *Auditing Quality* introduces the audit, a business tool for increasing compliance to quality standards. Few people outside the auditing field are aware of the value and function of either internal auditing or external auditing and the certification process. However, auditing is a powerful tool for quality assurance when it is properly integrated into the organization's management processes. We close Part 2 with a closer look at *Statistics for Quality* in Chapter 10. We learn when statistics do—and don't—apply, basic statistical concepts, and the most common statistical tools used in quality management. There's some good news for the nonmathematical: There are some tools based in statistics that do not require a statisticians expertise; we can apply them as nonmathematical managers.

By the end of Part 2: *Quality Essentials,* you will understand the most important tools and concepts of quality management. Then you will be ready to learn about the various schools of quality management that have developed these tools, why they emphasize different ones, and how they seek to help businesses deliver quality in the marketplace.

5

Key Quality Concepts

> *[A]s to the importance of obtaining the maximum output of each man and each machine, it is only through the adoption of modern scientific management that this great problem can be finally solved.*
> —Frederick Winslow Taylor, *Scientific Management,* 1911.

To manage quality, we have to agree on and understand some basic terms that define what we are working with. How do standards, error, and checking relate to one another? What are the various ways of preventing, finding, controlling, and eliminating errors? How do we decide how much error is acceptable? You will find answers to all of these questions in this chapter.

Requirements and Standards

The most basic idea of quality management is that we define good standards, then meet those standards through quality management processes. However, different notions of what we mean by a standard have created two very different perspectives. One view—where standards are seen as every requirement we have to meet except the customer specification—leads to a view that places quality as a peripheral issue, not centered on the customer. In this perspective, quality management becomes an added expense and a necessary evil of doing business. For example, in classic project management, requirements definition is part of scope management. We are scoping out what we are making, but we are not necessarily focused on how to make it good. Quality planning was defined as a separate process, and usually only said, "If this is what we are making, what government regulations must we comply with, and what industry standards would be helpful?" Then quality management made sure that these peripheral issues were addressed.

The other view—where the customer specification is included in the standard—makes quality management essential to all of our planning and work.

We are much better off including all requirements—including scope, time, and cost—in the definition of the responsibility of quality management. The second view—the meaning of the word Total in Total Quality Management—gives these benefits:

- *We can add customer value at little or no cost* by asking "What makes it good?" (quality) when we ask "What do you want?" (scope).
- *We can ensure a focus on value by ensuring a good specification.* We can apply quality definition, QA and QC methods to the process of creating the requirements specification to ensure that it is complete, consistent, and representative of customer needs and wants.
- *We can prevent the customer/quality divorce,* discussed in Chapter 4: *Quality for the Customer* by integrating quality management with product, service, and project management.

Defining Requirements

In Chapter 3: *Defining Quality,* we said that we can work with the customer to get a specification expressed as: a solution to a problem; a capability; functionality; or a detailed specification of the product or service. In addition, we have to add stakeholder requirements, technical requirements, regulations, and standards into the pot to create a complete specification.

Once we've done this, we have to turn the customer specification into a technical or engineering specification that can be used by internal workers and engineers. There are two parts to this process. One of them is *architecture,* which is the application of the ability to design something while seeing it several ways at once—and explaining it to others who can only see it one way or another, including the ability to work within constraints of different types. Architecture is beyond the scope of this book. The other, requirements definition, is a key step in quality management.

Requirements definition is the process of taking all of the requirements from different sources and combining them into a *requirements specification*—a single set of documents that gives everyone what they need to know about requirements in a format that they can read and understand. There will be an executive summary—a page or two in business language. The customers will get a summary and also a detailed, clear writeup of everything they asked for. But the bulk of the requirements specification will be a technical document for the team developing the product. It will have two components: a technical requirements specification, which can run to several hundred pages for something as simple as an automobile engine; and a *requirements tracing matrix* that links each requirement to its source (customer, stakeholder, or standard), and to the features of each component required to achieve the customer requirement. As we develop the product, the requirements tracing matrix is expanded to include our quality control plan, showing how each requirement is associated with various checks and tests. A full requirements tracing matrix is too big to fit into this book with everything else, so you can see an example at *www.qualitytechnology.com/QMD*.

DIFFERENT TYPES OF REQUIREMENTS

When we think of a requirement, we usually think of a definable or measurable aspect of a product. It is important to realize that this is only one type of technical requirement, and that there are several types of technical requirements and functional requirements, as well. For example, the functional requirement "satisfies my appetite, but isn't too big" will be measured in the size of the sandwich. But we may find that we specify two different technical requirements to satisfy that functional requirement for different people. That's what you see when you see a section of the menu called "children's menu" at a restaurant. As we will explain fully in Chapter 6: *Defining, Planning for, Controlling, Assuring, and Delivering Quality,* we can look at technical requirements in terms of whether they specify the input, process, or output of the whole product, or of a component.

- *An input requirement* applies when we check an input, such as a raw material or a subcomponent, provided by a vendor, before using it in our own process.
- *A process requirement* is a measure of a process as it is happening, not the measure of an element of a product. For example, if a method requires that an enamel finish be fired for 30 minutes at a temperature of 2500 degrees Fahrenheit, we will need a way to test if that was done. Or if a building

code requires that wiring was done by a certified electrician, we will need to be able to prove we used a certified electrician.

- *An output requirement* is what we usually think of as a definable, measurable feature of the output (product or component).

There are several important things to note about these three types of requirements:

- *Each type applies to components as well as to the whole product.* Using our ham-and-cheese example, an input requirement would be the use of a particular brand and grade of ham. A process requirement for each component might be that the ham be grilled, the bread be toasted, and the cheese be melted. An output requirement for a component would be that there be a certain total weight and thinness of slice of the ham.
- *Some component tests can only be tested before assembly.* It would be difficult to certify the total weight of cheese in the sandwich after the sandwich was made, but easy before the cheese was put into the sandwich.
- *Sometimes, requirements are interchangeable.* For example, we might be able to satisfy an output requirement—tastes spicy—by testing an input requirement—such as "used 1½ teaspoons of spicy mustard."
- *Process requirements generally require measurement or validation during the process and cannot be reliably obtained later.* For example, we can only know that our chef was wearing a hat and gloves—a health requirement—if we observe or confirm this while the sandwich is being made. It is possible to determine what happened in a process after the fact through investigation. We will discuss that topic in Chapter 8: *Auditing Quality,* where we introduce the issue of forensic investigation. But such investigation is less certain when records are not present, and is almost always much more time-consuming and expensive. The increased time and expense means that the only practical way to get good records of all the relevant check points on our processes is to gather data at the time.
- *Some requirements can only be measured by a destruct test.* The only way to know if a ham-and-cheese sandwich is really good is to take a few bites—so we'd better make some extras for testing!

If we are aware of all of these issues, we will be able to plan the checking and testing to ensure the quality of the results. Equally important, we can define the earliest and least expensive checks and tests required, so that we can deliver quality results at the lowest possible costs.

REQUIREMENTS, MEASUREMENT, TOLERANCES, AND ERROR

To do quality management, we have to fully define each requirement. What does a fully defined requirement look like?

- *A unique attribute is defined or measured.* For example, we might look at the type of bread, whether the bread is toasted, and the thickness of the slice. Each of these is a separate attribute, and we handle it separately.
- *Each attribute needs to be defined in a way that can be determined or measured.* For example, type of bread is defined (white, wheat, rye), but is not measured. Toasted could be either yes/no, or we could do a process measurement (length of time and toaster temperature), or a feature measurement (shade of color of the bread is within a defined range of medium to dark brown, neither too light, nor containing black.) The thickness of a slice would be measured.
- *Tolerances must be defined.* If a bread slice should be ½ inch thick, what is the allowable variation? Is anything between 0.4 inches and 0.6 inches acceptable? Or, we could have a range that doesn't have 0.5 in the middle, such as 0.40 inches to 0.65 inches. And how closely will we measure—in tenths of inches, or twentieths, or hundredths? If we don't assume that the slices are even, we would have to define multiple attributes, such as thickness of the thickest and thinnest locations on the slice. Of course, all of this is a bit silly for a slice of bread. But it would be deadly serious for the wheel of an aircraft.

Once a requirement is fully defined, naming the attribute, describing how it can be viewed, tested or measured, and defining tolerances, it is then possible to define error. An error is a single instance of one item where one attribute does not meet the determined requirement or is outside of tolerances when measured. Once we have defined what an error is, we can manage the errors to ensure conformance to requirements.

In addition to defining errors on an individual level, we can also evaluate their significance. In my ham sandwich, missing mustard and the accidental inclusion of a dead fly are both single errors, but one is certainly less significant than the other. The question of which errors to eliminate entirely, and which ones to minimize down to an acceptable level, is discussed in Chapter 14: *The Cost Of Quality*.

PLANNING FOR INDEPENDENT, COST-EFFECTIVE CHECKING

Clearly, a complex product will have many—perhaps thousands or even millions—of requirements that could be measured and checked. How do we plan to do all of that? There are two key issues: the independence of checking and testing, and cost-effectiveness.

The Independence of Checking and Testing

When a product feature is checked properly, it is checked to requirements. To ensure this happens, the checking process must be designed independently of the development process. Why? Because, in the process of development, we form an idea—a picture—of the requirements. And that picture might be wrong. Particularly, it might contain assumptions. The development of assumptions about requirements—thinking we understand the requirement correctly when we do not—is almost inevitable. If we then define the test with the same assumption, we are in big trouble.

What kind of trouble? We end up testing the product, but we're not testing it to the actual requirements. Instead, we are testing it to our interpretation of the requirements. The interpretation is wrong, and it is in both the product and the test. The test will say the product is good, but the product will fail to meet its requirements when it gets into the real world.

Although I've described this problem in terms of assumptions, it is actually larger than that. It includes any distortion of the specification, and assumptions are one kind of distortion. But the problem can occur due to measurement error in equipment, as well. A famous example is the Hubble Space Telescope. It's a wonderful instrument that has given us many years of good service. But as soon as it was placed in orbit, a problem was found. The mirror was not perfectly shaped. For a telescope to deliver an undistorted image, the mirror must be perfectly shaped to within ¼ of a wavelength of light. In spite of a great deal of care, the telescope wasn't up to snuff. There were many reasons for the problem, but the central fact was that one device was used for both cutting the mirror and measuring the cut. It was an impressive, and expensive, laser-controlled machine. But the lens in the machine that shaped the cutting and the measuring was a tiny bit less than perfect. When the mirror was cut, the shape wasn't perfect. And when the mirror was tested, the imperfect shape reflected back through the imperfect lens, which applied the exact reverse correction. Mirrors do that. So the telescope mirror passed its most demanding test, but was still flawed. And sending a corrective lens up into orbit was very, very expensive.

The same thing happens if we define our tests after we start planning and creating the product. The lack of independent creation of the tests causes us to let things slip through the tests that are not really up to spec.

There are two steps in preventing this problem:

- *Independent design.* All checks and tests should be designed independently of product development. If possible, the checks and tests should be defined by a separate person. If that is not feasible, the checks and tests should be defined first, before the product is designed.
- *Independent checking and testing.* As much as feasible, checking and testing should be carried out by people and equipment different from the people doing the work. People should definitely check their own work. That is valuable, but it is not enough. After a person thinks the job is done right, it should be checked independently to make sure it meets the specification, not his or her interpretation of the specification.

Cost Effectiveness

If we checked every relevant feature of every input, process, and output on every product in every possible way, the effort and cost would be staggering. Instead, we design our checks and tests to be sufficient to our purpose at the best possible cost. We control cost by choosing:

- *What to check and what to test.* Generally, checking is less expensive than testing, though with automation and computers, that is not always true. Also, checking and testing usually catch different types of errors.
- *When to check or test.* Usually, the earlier the better.
- *How to check or test.* Different methods have different cost per item. For example, I know a company that is developing the first nondestructive test for certain types of stress in metal. The innovation will radically reduce testing time and cost in certain applications.

Creating the plan for checking and testing requires technical knowledge of the field, awareness of cost, scheduling, business issues, analytic tools for evaluating test effectiveness and work flow, and good old common sense.

IMPROVING STANDARDS

Suppose that we have an existing product, and our standards and methods are all set up and working. Now, we find that, to win market share from the competition, we need to improve quality. How do we do that? How do we decide which new standard to follow, or which requirement to change? After we come up with a new set of requirements, we will adjust our methods so that we can produce to those requirements.

Obviously, we are required to follow current regulations. For example, the cooking temperature for ham changed a few years ago due to the development of more heat-resistant bacteria that cause disease. And we should know our industry standards, and evaluate updates, and apply them if they add value. Beyond that, there are three things we can do to improve quality standards: reevaluate customer requirements, use benchmarking, and adopt best practices.

Reevaluating Customer Requirements

Satisfying the customer always comes first. So, if the customer doesn't like our product anymore, or prefers someone else's, the best thing we can do is to find out why. Marketing surveys, focus groups, taste tests, and other approaches allow us to improve our requirements specification. It is important to realize that we must ask our customer about our price, marketing, sales, and customer service as well as about our product. Maybe our ham-and-cheese sandwich is still the best, but a major chain has stolen away customers with a lower price or a slick ad campaign. To illustrate with a real-world example, if a café starts losing business because Starbucks just opened down the street, the solution may not be better coffee: It may be better service or something else.

Benchmarking

A *benchmark* is a defined measure of productivity in comparison to something else. We can benchmark internally, seeking to maintain or improve performance, or we can try to find industry benchmarks, and compare ourselves to our competitors. The government often shares data that will allow for benchmarking; but within industry, the information is often proprietary. Sometimes, industry associations can provide information in support of benchmarks that we should achieve. We should always evaluate them closely to be sure that the benchmark is appropriate and realistic in our work environment. For example, if we are using older equipment, we might not be able to achieve an industry average rate of production. Also, we should make sure that achieving that benchmark increases or at least maintains customer quality while lowering cost. There is no point achieving a benchmark if it means losing customers or losing dollars.

Best Practices

Information about solid, measurable benchmarks is hard to obtain and harder to fit into your unique situation. However, there is another approach that is easier, more flexible, and more suited to unique and innovative situations. Developing and using *best practices* is a powerful improvement method. A best practice is simply the best way to do a repeating process at your organization. Best practices:

- Must be defined, written down, and repeatable.
- Must meet or exceed applicable standards and regulations.
- May be derived from general standards, industry standards, industry best practices, research into activities of other companies or industries, consultant expertise, or internal research.
- Must be adapted and optimized for your organization.
- Are found, implemented, and improved over time.
- May be very technical and specific, or be broad and adaptable. If they are broad and adaptable, there should be guidelines for effective adaptation included in the practice.
- Must be more effective than any other available method, so that they are truly best.

Best practices begin with humility, with letting go of thinking, "I know how this should be done." The next step is genuine research: Who knows the best way to do this? The next is innovative adaptation: How can we do it even better here? When an organization takes a best practices approach, it will need to support that approach. The company should implement methods and systems of storing, organizing, reviewing, improving, teaching, and adapting best practices.

Checking

Checking is a general term we will use for any activity that compares the product or service against requirements, or that compares the technical process of creating the product or service against requirements. Each act of checking determines if a given feature is acceptable—within tolerances—or is an error outside of tolerances.

Quality management terms are sometimes confusing and mysterious because they were created at different times and places, different words were used for the same thing, and meanings sometimes overlap. There are many kinds of checking, but what they all have in common is that they compare the reality of the product or service—or the process that makes it—against requirements or standards. With this definition of checking, we can clarify three other very important terms.

- *Quality Control (QC)* has two meanings. The narrow meaning refers to statistical quality control, where we test a small sample of the entire product batch and extrapolate to define qualities of the entire batch. In the broader use, *quality control* is synonymous with checking. It refers to all activities of review, inspection, and testing of the product or its technical process, with or without sampling and statistics.
- *Quality Assurance (QA)* includes quality activities outside the realm of checking and quality control. QA includes cross-departmental communication about quality, communication with vendors, redesign of the product or process to prevent error, and a variety of audit processes to make sure that work and management are being done to standards or in accordance with best practices.
- *Quality Planning (QP)* is a newer term from the 1990s. It includes all early efforts to plan how error will be prevented, how quality will be managed, and some of the design activities of QA.

DIFFERENT TYPES OF CHECKING

There are many types of checking suitable to different circumstances. We can break them down into three basic types: review, inspection, and testing. We review written documents. We inspect or test components, products, or services.

In Checking, the Sooner, the Better

Before we look at each type of checking, we should grasp one essential principle of quality management. The sooner—the earlier in the process—we eliminate error, the better.

The best choice is that, with good requirements specification, architecture, and design, we prevent the error altogether. So error prevention is the least expensive option. In order of time and cost, here are six ways to deal with error:

- *Do it right the first time.* Put top-notch effort into quality definition, planning, and design, so that the product or service is as free of error as possible from the beginning.

- *Catch the errors in early reviews of the plans.* We should invest a lot of our quality management effort in review—ideally, close, structured review—of plans and design documents.

- *Doing good production work.* If we work as a team to make sure that inputs from vendors are checked, equipment is working well, and people are following procedures, we minimize error and waste in production.

- *Checking production work.* This would include all forms of inspection and testing, with correction of the defect or scrapping of the component to prevent the defect from reaching the customer.

- *Letting the customer receive the error, and then doing a good job of fixing it.* Here, the customer has to deal with the frustration of the error, but we do a good job of helping with the cost through warranties, service plans, and affordable, high-quality customer service.

- *Letting the customer receive the error, and then not providing good customer support.* In this case, the customer pays the price, and we almost certainly lose the customer. We also risk loss of reputation and legal action against us.

One way to fully appreciate the above list is to realize that, for every error, one of these six things will happen. We will address this fully when we discuss the *Cost of Quality* in Chapter 14.

Another way is to realize that these six errors fall into three sequential stages—planning, development of the product or service, and delivery to the customer. Many studies across all industries have demonstrated that there is a cost and time ratio for planning:development:delivery of 1:10:100. This is called the 1:10:100 rule, and it states that each error will cost ten times more to fix in development than it would to fix in planning, and 100 times more if the error actually reaches the customer. Some experts, most notably Dr. Harold Kerzner, have discovered much higher ratios. Dr. Kerzner cites a client who found that, in a five stage project life cycle, the ratio was 1:5:25:100:1000. The cost and time ratios and consequences of errors of the 1:10:100 rule are laid out in Table 5-1.

Let's take a closer look at the three basic methods of checking:

- *Review* is the process of comparing a document, such as a requirements specification or a design plan to standards or requirements that govern the process or results required of that document. Reviews can be highly formal and strict, or they can be loose and informal. Generally, the higher cost of the time effort spent in a formal review pays off, because it means we catch more errors earlier in the process.

Table 5-1. Cost and time ratios and consequences of errors of the 1:10:100 rule.

Stage	Do good work to prevent error	Check, catch, and correct error	1:10:100 ratio	Time consequences	Cost consequences	Customer consequences
Plan	Do it right the first time		1	Best option	Best option	Least error; most satisfaction
		Catch errors in early reviews of the plans	1	May delay beginning of development	Relatively low cost, errors are easier to fix in plans than while working	Least error; most satisfaction
Develop	Doing good production work		10	Keeps production and delivery on time	Ten times more expensive per error, but continuous improvement here saves a lot of money	Customer will not see error, but increased cost may be passed on to customer
		Checking production work	10	May delay product delivery	Expenses can balloon at the end of development, cost and reducing return on investment	Customer will probably get product late, and cost increase will probably be passed on
Deliver	Customer receives error, we fix well		100	Time to repair for customer	We have to maintain a larger customer service team and repair facility, loss of future business	Hassle for customer, cost may go to company or customer—customer likely to be dissatisfied
		Customer receives error, we do not fix well	100	Time for multiple repair efforts	Very large customer support cost consequences, loss of future business, loss of repair income, loss of reputation, legal costs	Hassle, cost, and customer dissatisfaction with product and with company

- *Inspection* is the act of examining an attribute of a product, service, or component and comparing it to its specification. Some comparisons are discrete, such as "Did a red, blue, or black t-shirt go into the box for the customer?" while others involve measurement. Where measurement is involved, we determine if the attribute is within specified tolerances. *Statistical quality control* is a special case of inspection where we test only a sample of the product and extrapolate to statements about the entire batch of the product using statistical methods.
- *Testing* is the process of actually doing something with a product, service, or component and seeing what happens. Key issues in testing include the design of experiments, the cost of testing, and the type of test. Tests should be designed to check the maximum number of features at the lowest costs. *Destruct tests* are tests that check a feature, but destroy the product in doing so. Clearly, destruct tests can only be used on prototypes or samples of our final product, not on every item we were going to sell!

QUICK QUALITY TIPS

But Did You Fix It?
Checking isn't enough. After checking, we need to either discard the flawed item, or do rework to fix it. If we do rework, we then have to check it again to make sure it passes the test this time, and still passes all previous tests. All too often, when we're in a rush or under pressure to be productive—to deliver quantity on time—we do see errors in checking, but we let them slip through anyway. The cost of this is huge in dollars, in lost sales, and even more in the company's reputation with its customers and its employees.

Using the Information from Checking

When we perform a check, the biggest mistake we can make is to ignore the results or slip up, allowing the error to move forward in the production process or even to the customer. Instead, we can use the information to eliminate the error from the product, and we can also use the information to prevent future occurrences of the same and similar errors.

ELIMINATING ERRORS FROM THE PRODUCT

The traditional use of inspection before Shewhart and the earliest uses of quality control focused on using inspection or statistical quality control to eliminate errors from the product. When an error was found, the focus was on the particular instance of the error, and we had these options:

- *Allow the error to pass through.* Due to the 1:10:100 rule, this is almost always a bad idea. But it happens due to failure to communicate, and also due to pressure to increase productivity or reduce costs.
- *Rework.* In rework, we fix the error in the single product, then make sure the reworked product or component passes all tests.
- *Scrap.* We can throw away the product or component.

Traditionally, the choice between rework and scrap was based on cost. If the cost of the materials plus work done to date was high, then rework is better. But if the cost of rework was high, then scrapping is better. On the shop floor, the choices are more complicated. If the item is scrapped, can its raw materials or components be reused or sold? Or, can the item be sold for another purpose. For example, I have a keychain—a memento from a trip to a state park—that is a carabiner—a clip used for rock climbing. It wasn't good enough to sell for use in rock climbing, but could be sold at a lower price as a keychain.

No matter which option we choose, the cost of inspection plus follow-through to deal with the error is high. Often, this cost can completely eliminate the profit margin for a manufacturing production operation or assembly line. This high cost led to the discovery of techniques to prevent error, including statistical process control, quality assurance, Total Quality Management, and process re-engineering.

PREVENTING FUTURE ERRORS

Taylor introduced the idea of using what we learn from experiments to change processes in ways that reduce error, and Shewhart made the methods more available by introducing PDCA. Shewhart also introduced *statistical process control,* which had two main elements. One was cost reduction through examination of a sample, rather than an entire batch of parts. The other was the use of the information about errors to go back and change the process so the error won't happen again. In practical, daily life, this seems obvious. If we are baking several batches of cookies, and the first batch comes out overdone, we go back and either lower the oven temperature or shorten the baking time. Conceptually, this is called feedback for course correction. Practically, however, many hassles and barriers prevent this from happening in companies where people are working together. There are political and bureaucratic barriers such as union resistance to changing work rules, elements of ego, such as the statement "don't tell me how to do my job," elements of unclear thinking, such as, "how do we really know that this will be better, and won't lead to other problems," and elements of simple miscommunication.

We avoid this through the actual application of scientific management. In science, ego bows to truth and to what is best as determined by observation and experimentation. Failed communication is traced and repaired. The first time this was done on a large scale was in the Total Quality Management movement, beginning in the 1950s in Japan and in the 1980s in the U.S. Organizations that adopt these methods open the door to continuous improvement, called *kaizen* in Japan, where it was first developed. In most workplaces, standards are haphazard, poorly understood, or, at best, something to strive to meet. In the company or organization that embraces continuous improvement, standards are clear and available, and workers develop them and are trained to meet them consistently. Even more important, in the words of Dr. Masaaki Imai, founder of the Kaizen Institute, "today's standard is the worst possible way to do a particular job." That is, we always deliver by using at minimum the standard method that we have. And we are always asking, "How can we do it better?"

When we find a way of doing it better, the company supports the experiment to see whether we have found a new best practice, and, if we have, streamlines the process of making sure that the standard is updated, all workers retrained, and the new best practice is adopted and followed until it is replaced by something better.

Quality Management as Error Management

We have already discussed how increasing quality is the same as reducing the instances of error. To the extent that our specification is error-free and accurately represents what the customer wants, and we deliver a result meeting or exceeding that specification, we deliver quality. There are three key ideas inherent in that statement: the relationship between process and product; idea of the extent of error; and the trade-off inherent in exceeding a specification. These ideas are discussed in the next three sections.

PROCESS AND PRODUCT

Does every error in process result in an error in the product? That is, does an error in how we do the work create an error in what we deliver? The simple answer is, if we define our processes in a relevant way, yes. But often, it isn't that simple. The standards for our process may require that we do things a certain way but, when we deviate from that method, error may not result. And sometimes, even though we follow standard procedure, we get errors anyway.

Consider the standard that requires that household wiring be done by a certified electrician. What if it isn't, but the electrician, although not certified, does a good job? There may be no error in the product. But can we know after the wall is closed up? Several things could happen. A state inspector might delay construction or require breaking into the wall for inspection because the work did not meet the code requirement—a process requirement—that it be done by a certified electrician. Or perhaps the error in process will be overlooked, and we will just hope that the work was done well. Perhaps the error won't be re-inspected, because it is too late to do that affordably, but we will be fined. Perhaps the customer will refuse to buy the house because of the risk—real or perceived—arising from failing to comply with the standard.

Equally, a certified electrician could do the work, so that there is no defined error in process, but he could make a technical mistake that could result in an error.

It is a universal principle that good work leads to good results. Part of our job as quality managers is to have the right definition of "good." In quality management, we should do our best to align required processes so that our definition of "good work" is "that which leads to good results."

DEFINING ERROR AND ITS CONSEQUENCES

When we perform an inspection or test, we see the *symptom* of an error, the evidence that it exists. A full description of the error also includes:

- *The cause,* or reason the error happened. This is a description of the variation of input or process that led to the error. There may be multiple causes and relevant surrounding conditions.
- *The root cause* which is the deeper reason the error, and this type of error, occurs. For example, a cause of a dent might be dropping a part, and the root cause might be failure to train new personnel in proper handling of materials.
- *A consequence,* which is what would happen if the error went undetected.

- *A corrective action,* such as rework or scrapping, that takes care of this error.
- *An adjustment to process* to prevent the error from continuing to occur in the current process of production.
- *A permanent preventative solution,* which is the solution to the root cause, so that this error and ones like it will not happen again in this process or any similar process within the company.

The consequences of error, and the cost of those consequences, are the primary business justification for quality management, as we shall see in Chapter 14: *The Cost Of Quality.* Root cause analysis and permanent preventative solutions are the basis of continuous improvement in TQM, *gemba kaizen,* and an element of other quality management methods that we will look at in Part 3.

HOW ERROR-FREE DO WE WANT TO BE?

It is easy to say that we want to eliminate all errors and deliver products and services with zero defects to the customer. It is very hard to achieve. And it may not be the best thing to do, either.

There are certain environments where being completely error-free as often as possible is essential. These include:

- *Where life is at risk.* Aerospace, military, and medical applications are leaders in the zero-defect movement.
- *Where the chances of finding the error later are very low, and the costs of failure high.* The U.S. National Aeronautics and Space Administration, NASA, while a source of many excellent quality methods, is, unfortunately, also a very visible example of this type of error. If there is something wrong with the engineering or programming of a landing mechanism, if it isn't detected, it won't be discovered until the spacecraft crashes.
- *Where the cost of replacement on location is high.* Here, NASA's Hubble Space Telescope, where the cost of building a corrective lens for a misshapen mirror was small, but the cost of flying into space to install the lens was very, very high is a clear example. But the same is true for anything that is mass-produced, where we have to either recall or go out and repair every product after it is sold.
- *Where reputation is at stake.* The highest caliber artists—including musicians, dancers, martial artists, TV and movie producers, and others—strive for error-free results both out of a desire for perfection and also because a single error can set back or even ruin a career.

Those of us interested in best practices and continuous improvement can learn a great deal from those whose jobs require error-free delivery. As my colleague Dan Millman puts it in his book *No Ordinary Moments,* although our life is not on the line at every moment, "every moment, the quality of your life is on the line."

In business, though, we may determine that there is an acceptable level of error. Generally, if we can eliminate error in planning, that almost always pays off. After that, there is a trade-off between the cost finding and fixing errors, versus the cost of warranty repair when the item fails for the customer. Determining how much error to allow is difficult because, of course, we never know an error is present until we find it.

For example, imagine that you are asked to make sure that a manuscript is free of typos. On the first pass, you make fifty corrections. On the next pass, ten more. On the next pass, only three. On the fourth pass, you make only one correction. So, each pass—and all of them are equally costly—finds fewer errors. What if the next pass finds no errors? Does that mean that there are none? Or does it mean we need to keep looking to find the errors you missed, or that we need someone else with a different perspective or skill set? How can we ever know that something is error-free?

The short answer is: we can't. And the fewer errors there are remaining, the more it will cost to find each one. Solutions to this problem are discussed in Chapters 8, 10, and 14.

Why Errors Matter: A Systems Perspective

Before we close the chapter, we need to take one more look at the issue of errors and their consequences. We must address an issue that has troubled many people for centuries: Why are there some errors that don't matter much, and others that have huge, catastrophic consequences? Is there any way to know which errors are most critical?

Yes, there is, and the method is called systems engineering. If we view a product, service, or process as a *system,* we see it as a set of parts that work together for some purpose within a larger system. Those parts, in turn, might be systems that we call sub-systems. For example, a person might have a purpose in life. The body is part of that person. The body is a system made up of sub-systems, one of which is the circulatory system. The circulatory system, which has the purpose of transporting blood to provide oxygen and nutrition, has a component called the heart, and two others called lungs. If the heart fails, the circulatory system fails, the body fails, and the person can no longer fulfill his or her purpose. But if one lung fails, the other is enough to do most of the work, and the circulatory system and body can keep going. John Wayne made movies for a decade after he had one lung removed. He was limited—he needed oxygen after performing some of his stunts—but he was still working.

When we understand our product as a system, we can trace the consequences of any error or defect more completely. Applications of systems thinking to quality management could fill several books. Here is a short introduction to some of the ideas that are most valuable and easiest to apply:

- *Tracing cause and effect* through a system is an essential first step.
- *Understanding causal loops* where a cause creates an effect which creates the cause happening again—the cycle of chicken producing egg producing chicken, over and over—is essential. Stable loops can leave us either always getting what we want—like a thermostat controlling a heater—or never getting what we want—like a bad marriage where talking about what's bothering us makes us angry, so we have more that is bothering us. Unstable loops fly out of control. If I get angry when you get angry, we both start yelling and have a fight. If I don't talk about problems and you don't talk to me when I'm not talking about my problems, communication drops to zero. Machines with control systems have exactly the same problems.
- *Systems can run in serial or parallel.* Serial systems, like a chain, fail if one link fails. Parallel systems keep running if one component fails.
- *System robustness can be defined and measured.* Robustness is a measure of a system's ability to endure or recover from damage or injury. It is more than simple strength. Issues include strength, flexibility, redundancy, self-correction, self-repair, and self-healing.
- *Redundancy—two parts doing the same job—makes systems more robust.* That's why we can keep living if we have only one lung or kidney. I once advised a medium sized company that wanted to work 16 hours a day, 5 days a week, "anything you need to have working all the time, have two of them." Redundancy is often an inexpensive solution to problems of reliability.
- *Some components are not part of crucial systems, and others are.* On a car, we can drive with a dented body panel. But if anything is wrong with the drive train or the braking system, the car is not safe to drive.

We can also look at our process—our methods of producing or delivering a product or service—as a system. This allows us to trace backwards from symptoms—defects or errors—to causes and root causes, and engage in process improvement.

Conclusion: Understand, Then Improve

While I was writing this chapter, I called my dad. I told him that I'd found that almost all problems in quality management had been solved by Taylor by 1911, and that the only really new thing since then was Shewhart's inclusion of statistics for quality control in 1924. He replied, "We've solved the problem in theory, but not in practice." And he's right. That is what makes this chapter so important. If you are using any quality technique—QA or QC or quality engineering or auditing—or any method—Six Sigma, TQM, ISO 9000—and you are having problems, then the solution is most likely somewhere in the basic understanding of what your company is doing. Returning to the basic definitions in this chapter and the core methods in the rest of Part 2 will help you figure out and fix what isn't working in your quality management methods. And that sort of work is essential. An error in our quality methods almost always means systemic problems that are hard to find and fix, and multiple problems in our products.

Companies change, and they tend to spiral. The downward spiral begins with insufficient attention to quality or a poor choice of quality management methods. It leads to more errors, less revenue, higher costs, and fewer resources to fix problems, so that the cycle continues downward until there is a crisis. The other option is the upward spiral of quality. We eliminate defects, reduce costs, put standards in place, meet those standards, then engage in a cycle of continuous improvement, a never-ending upward spiral of value for us and for our customers.

Q-Ball Quiz for Chapter 5

1. Which of these is the most effective way to use requirements?
    (a) Separate standards from customer, technical, and stakeholder requirements, and deal with each one independently.
    (b) In business, allow the customer to define the standards you must meet.
    (c) Define a set of requirements that includes customer requirements, stakeholder requirements, technical requirements, and applicable standards and regulations, then strive to meet it.
    (d) Constantly seek to improve and change requirements throughout product development.

2. Which of these is *not* one of the three types of requirements?
   (a) an output requirement
   (b) a statistical requirement
   (c) an process requirement
   (d) an input requirement

3. Which of these is *not* true of the 1:10:100 rule?
   (a) It describes the ratio time and cost for error elimination across planning:development:delivery.
   (b) It has been found to apply in many industries and situations.
   (c) Lower ratios have been found in certain cases.
   (d) Higher ratios have been found in certain cases.

4. Which of the following is *not* a legitimate option when a single defect is found in inspection or testing of a single instance of product?
   (a) Allowing the product to go to the customer with the defect.
   (b) Reworking the product to eliminate the defect, then making sure that the product is error free before delivery.
   (c) Selling the product for a different use with less demanding requirements.
   (d) Scrapping the product.
   (e) All of these are legitimate options

5. All of the following methods have been used by businesses to deal with known errors in products. Which one is *least effective* for the business in the long run?
   (a) When an error is found, fix it or scrap the product with the error.
   (b) When an error is found, trace the effects forward and evaluate the consequences.
   (c) When an error is found, trace the causes backwards to determine causes and root causes.
   (d) Define a solution to the error, and inform customer service representatives, so that the company can make money from repair bills and service contracts.

6

Defining, Planning for, Controlling, Assuring, and Delivering Quality

*Scientific management fundamentally consists of
certain broad general principles, a certain philosophy,
which can be applied in many ways.*
—Frederick Winslow Taylor, *Scientific Management*, 1911.

Now that we have covered the key concepts of quality and quality management, we can look at how to put together a simple framework for successful quality management in your business. This overarching model with five stages—quality definition, quality planning, quality control (QC), quality assurance (QA), and quality delivery (customer delight)—includes all of the methods used by the movements discussed in Part 3. In fact, we can define many of the differences in these move-ments by which of these stages they emphasize. The stages are only summarized here, as we provide more detail about each of them in other chapters, as noted.

Quality: A Business Perspective

Quality management seeks to change business processes to assure and increase value for both the customers and for the business itself. But a business is a sys-tem, and a change in any one process has complex consequences. Usually, a process is changed not just once but repeatedly in a series of progressive refine-ments. The first changes bring big results, but, as time goes on, we have to spend more time and money to get useful results. Once we've fixed the big, obvious problems, one of two things will happen: either changes will become incremen-tal, resulting in smaller benefits, or the company will hit one bottleneck—one area that it doesn't see or isn't willing to change—that is the source or *root cause* of many problems earlier thought to have stemmed from other causes that by now have been removed. When either of these start to happen, executives can become disenchanted with quality management programs or continuous improvement, and start to look elsewhere for magic pills, quick fixes, and big-ger bang for the buck in improvements.

In addition, in quality management, the ratio of improvement effort to ben-efits varies greatly. Sometimes, a single change in a process will bring a lot of benefit. Other times, three or four changes in a process have to come together to produce one change in results. We can understand this best if we visualize our business as a series of pipes containing products, information, and money. If one main pipe is broken in just one place, then fixing it will make a big dif-ference. If the same pipe is clogged in five different spots, then we have to clear all five clogs before things get moving and we see increases in flow or produc-tivity. Also, once we've fixed all the pipes and cleared out the main ones, we will see less improvement in overall flow as we get down into the details and clear out small pipes that are only partly clogged. If results are incrementally lower over time, why should a company stick with quality management? There are three reasons:

- When we don't spiral up towards more quality, we quickly spiral down towards less. It is hard, if not impossible, to stay on a plateau. Maintaining quality in a world of human error, mechanical failure, and general change requires constant checking. That constant checking—if we do it—naturally leads to some degree of improvement. Without that constant checking, quality is sure to degrade.

- If our competitors keep the focus on quality in an effective way, we will fall behind fast—particularly in market share.
- A steady commitment to quality solves other problems, such as customer retention and employee retention, reducing cost of sales and cost of operations.

Quality: A Process Flow Perspective

The idea of seeing a business as a bunch of flows through pipes is not a new one. Chemical engineering developed a method called *flow diagramming* for chemical plants, and that led to *data flow diagramming* for business by the 1950s. We also talk about work flow, information flow, and cash flow. When things are flowing, it's good for business.

In a chemical factory, stuff flows through vats and pipes. What carries the flow of work in a company? Processes! In a factory, a process is how a product is built. And we can define each process as having seven elements, as shown in Figure 6-1.

We can think of each task as having three core elements, and four ancillary elements. The major elements are:

- *Inputs,* which are the ingredients, raw materials, or components that go into a process and become part of the output.
- *Process,* the activity of transforming inputs to make outputs—the work.
- *Outputs,* the end results of a task, such as a component or a finished product.

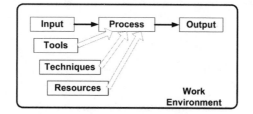

Fig. 6-1.   The input-process-output model of a process or task.

The additional, minor elements are:

- *Tools* or equipment, which are used for the task, but not used up.
- *Resources,* including disposable items (such as cleaning supplies), and our effort, which are used up in the process but do not get included in the product.
- *Techniques,* the instructions for the work process.
- *The work environment,* the space and conditions within which the work is being done.

Of course, each product may be built using many tasks—perhaps thousands or even millions. Tasks are linked because the output of one task is the input of another, until we can link the suppliers through all the tasks to the customer. This is called the Supplier-Input-Process-Output-Customer model, and is illustrated in Figure 6-2.

Of course, a single product or company has many such chains that link all suppliers through many processes to all customers.

We can map the five stages of our quality management framework to the SIPOC model as follows:

- *Quality definition* comes before the definition of processes.
- *Quality planning* includes defining what processes are required to deliver the product to meet or exceed specifications, putting them in order by linking outputs of one process to inputs of the next, and then defining all seven aspects of each process with requirements and tolerances on all key variables, so that we can consistently produce all outputs of all processes to specification.
- *Quality control* in the broad sense including all forms of checking, ensures that outputs and processes meet requirements, that defective output is

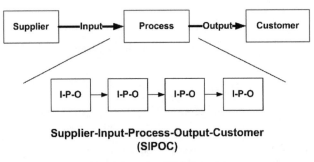

Supplier-Input-Process-Output-Customer
(SIPOC)

Fig. 6-2.   The SIPOC model.

reworked or scrapped, and that all seven aspects of processes are adjusted and restored to work within tolerances.

- *Quality assurance* includes activities to evaluate and improve processes, re-engineer work to eliminate unnecessary processes or steps, ensure effective communication and mutual understanding throughout the SIPOC chain, and auditing and review to ensure all processes are maintained to standard and improved.
- *Delivering quality* means carrying the SIPOC chain all the way through to the customer's receipt of the product or service, to the customer's perception that he or she has indeed received value and quality in the product, service, and contact with the company.

ALL THE ANGLES

Does Quality Flow Through Your Company?
Take some time to sketch out a rough flow of the work of your company. Focus on movement of products and information, not on the reporting hierarchy. Then look over the five stages of quality management. Then write down your answers to these questions:

- Which departments have the most, and the fewest, problems?
- Which vendors provide the least, and the most, reliable raw materials and components?
- Which customers are most and least satisfied?
- Of the five stages of quality management, which one do you do best?
- Of the five stages of quality management, which one could most use improvement?
- Does your company, or one particular area, need a quality improvement program?
- If you could go make one change to fix one thing today, what would that be?

You might return to this exercise two more times: when you finish this chapter, and when you finish *Quality Management Demystified.*

Defining Quality: Requirements Elicitation

We have already learned a lot about requirements elicitation in Chapters 3 and 4. Now, let's go a little deeper by looking at effective approaches to requirements elicitation and at the qualities of a good requirements specification.

Here are some excellent practices for obtaining good requirements:

- *Define your goal clearly at the start.* Are you seeking to define a new product? Are you seeking to learn about what your customers do and don't like about your current product? Are you seeking to compare customer opinions of your product or service and your competitors'? Are you seeking to determine what specific changes to your product or service the customers most want?
- *Make it interactive.* We learn more by letting customers try out or taste or play with a sample or prototype than we do by asking questions. We want the customers focused on the product, not on us. If a prototype or sample isn't possible, then we should use pictures, charts, and diagrams.
- *Record everything.* If possible, videotape or audiotape the sessions. If not, have two notetakers so you lose as little as possible.
- *Use industry best practices,* such as focus groups and structure requirements elicitation methods.
- *Learn and use good survey design.* Good surveys are harder to make than you would think. McGraw-Hill's *Business Statistics Demystified* is a great place to start.
- *Study your results.* Don't just gather a lot of data and ignore it. Put it all together and learn what you need to know. *Quality Management Demystified* will help with many analytic tools, the most important of which is plan, do, check, act (PDCA).
- *Check and test your results.*If you have a limited set of customers, or customer representatives (such as a marketing department), have them check and improve what you come up with from the sessions before it goes final. Otherwise, use multiple methods, such as a survey, a focus group, and a limited pilot product launch before you go into full production.

After eliciting requirements, we will need to organize them into a clear, useful requirements specification. The best description I've found for what makes a good requirements document is from the Institute of Electrical and Electronic Engineers, in their Standard 830-1993, "IEEE Recommended Practice for Software Requirements Specifications," summarized here in Table 6-1.

Planning for Quality

Quality planning includes all of the work we do to organize and layout a plan for all five stages of our quality work. Some of it begins even before we define product quality through requirements elicitation. After we have the requirements specification, we do a lot more quality planning. We define all of the reviews, inspections, and tests. We define our approach to QA, QC, rework and scrapping, process improvement, and delivering customer delight. Of course, we may have standard practices we always use, in which case we adapt them for this par-

Table 6-1.   Some characteristics of a good requirements specification.

Some Characteristics of a Good Requirements Specification Adapted from IEEE standard 830-1993 and *Creating a Software Engineering Culture* by Karl E. Wiegers	
Characteristic	**Description**
Complete	Nothing is missing, all attributes relevant to customer satisfaction are included, defined, and given tolerances.
Consistent	The specification contains no internal contradictions.
Correct	The specification accurately reflects customers' and stakeholders' wants and needs.
Feasible	Delivering to the specification is possible with technology that is available, can be obtained, or can be developed. Delivering to the specification is possible within time, cost, and other constraints.
Modifiable	The specification is designed so that future changes can be made in a defined, practical, traceable way.
Necessary	Each requirement adds value for the customer.
Prioritized	Requirements are ranked as to how essential it is to include each in the book. A group at the top may be listed as required, and then optional ones listed below that, in priority order.
Testable	Each requirement must be defined in a way that will allow for one or more tests of either process or product that will ensure conformance and detect error.
Traceable	Each element is uniquely identified so that its origin and purpose can be traced to ensure that it is necessary, appropriate, and accurate. This usually means assigning a number or code to each requirement that doesn't change, and then adding codes to indicate changes to a requirement and giving each new requirement its own code or number.
Unambiguous	Each requirement has only one possible interpretation.

ticular project, service, or product. We also need to make sure that our practices are up to date, conforming to current versions of relevant standards and regulations. And we may want to research best practices, as well.

Quality planning doesn't end there. As we develop and deliver the product, we can use any information obtained about quality and quality management from any process to improve the way we do quality work. When we do, that will require additional planning. For example, suppose a team member suggests a way that we can sell items that were being scrapped to a different, less demanding market. In that case, we need to do lots of planning work: We define the new market, see if it is really viable, then define tests for the scrap to make sure they meet the lower requirements of that market. When we make use of what we learn as we go by feeding that knowledge into ongoing quality planning and then implementing those plans, we accelerate the continuous improvement cycle and add value for the company.

Checking: Quality Control and Inspection

Under the broad term *checking,* I am including all of what was classically called inspection as well as statistical quality control. The two have a lot in common. Both have the goal of conformance to specifications. Both require clear specification of each attribute, so it is definable and subject to a process that determines if the result matches the requirement or falls within tolerances. Both can use review of documents, inspection of attributes, or testing of components as the method by which information is gathered. The results of both lead directly to a choice of allowing the error through, rework with retesting, or scrapping. Both also feed information back into the control of the manufacturing process, so that future, similar problems can be prevented by bringing the process under control, and by performing root cause analysis and applying a permanent preventative solution.

STATISTICAL QUALITY CONTROL

The only difference between statistical quality control and inspection is that, in inspection, we examine every single product. When we have thousands or millions of products made to the same specification, we can save time and money by applying Shewhart's statistical techniques and doing statistical quality control.

If we can show that a sample of the product falls within a narrower tolerance, called the control limits, and certain other requirements are also met, then we can be confident that the whole batch fits within the customer's wider tolerance, called the specification limits. This is explained in more detail in Chapter 10: *Statistics for Quality.*

CLARIFYING IDEAS ABOUT INSPECTION

In some circles, inspection has gotten a bad name. This is a result of misunder-standings of things that a number of quality gurus, including W. Edwards Deming, have said. Deming did say, "Quality comes not from inspection, but from improvement of the process," and "The old way: Inspect bad quality out; The new way: Build good quality in." (Walton, *The Deming Management Method,* p. 60) Unfortunately, although he also said, "eliminate slogans," (ibid., p. 76), people have turned his comments about inspection into a slogan, and then misunderstood and misapplied it.

Deming's point was not that we should stop performing inspections. It was that we should use the information from inspections to do more than just move the one product into the scrap pile or rework line. We should use information from inspec-tion to improve the way we do things so that we can build quality in. Inspection is essential to Deming's Total Quality Management because, by detecting errors, we gather information for the root cause analysis that will allow us to eliminate them.

The second confusion about inspection is that some people discourage it because they think it has been replaced by statistical quality control. It has, but only in very limited circumstances. We can only apply statistical methods when we have large quantities of identical components or products—generally tens of thousands or more—and we can make precise measurements to match against specifications and tolerances. As a result, statistical quality control is effectively limited to manufacturing, automated services, and some very high-volume ser-vice industries. Elsewhere, we must rely on inspection.

The third misunderstanding regarding inspection is that some people think that inspection implies inspection by people, by human beings, rather than by machines. There are limits to the effectiveness of inspection by human beings. However, we are now finding ways to develop computer-controlled sensors and robotic devices that perform inspections. In electronics manufacturing, for instance, there are robots that can manipulate a circuit board and test every circuit on every board, so that inspection is once again replacing statistical qual-ity control. While human inspection cannot eliminate all errors, automated or robotic inspection is part of the wave of the future.

Quality Assurance

Quality assurance (QA) developed in North America while TQM was developing in Japan. QA focused on solving quality problems, rather than living with rework and scrapping. The major difference between QA and TQM is that QA was usually performed at an engineering or management level with little executive support. Also, QA tended to gather information by auditing after the fact, which meant that it didn't bring in rapid benefits the way that methods that focus on the earliest parts of the process can. QA focused on production—the ratio of 10 in the 1:10:100 rule—more than on planning and requirements definition, where there is more to be gained.

Nonetheless, QA does offer some distinct and useful ideas:

- *Product re-engineering.* This is where we examine failure points of the product, then redesign it to have fewer failure points. For example, if solder joints often fail in testing or too soon after delivery to the customer, we might design the product to have fewer parts so that there are fewer solder joints, or we might use a different method for joining parts.
- *Process re-engineering.* This is where we find better ways to do the work, reducing errors.
- *Evaluation of customer satisfaction and customer service information.* QC only had information from inside the factory. QA gathered information from customers, service companies, and the company's own repairmen, and used that to identify problems.
- *Communication across departments.* QA experts saw that many quality problems happened when there was an unclear definition of requirements across departments. If your output is my input, but we never talk to one another, we're going to have quality problems.
- *Communication with vendors.* Similarly, bringing internal and vendor engineers and managers together to resolve quality problems across the supply chain was a part of QA.

QA offered many good ideas and added value, but did not produce the transformative results that TQM did. There were two reasons for this, which I call "too little" and "too late." By too little, I mean that QA had too little clout. Engineers and some managers saw its value, but they were unable to convince upper management. (Deming saw this problem, but was unable to solve it in the United States) As a result, when productivity demands increased, people working on QA were told to "quit talking and get back to work." QA was actively discouraged for another reason, as well. Until the difficult economic times of the 1970s, Henry Ford's methods were unchallenged in North America. The assembly line focused on productivity and efficiency, not on quality. Worse, shops were run by Theory Y managers, and workers were assumed to be slackers. Disrespected workers built strong unions that were very ready to strike. In this environment of conflict, the quiet voice of the QA engineers and managers who saw a better way—a way of cooperative improvement—was lost.

By too late, I mean that QA focused primarily on audits and defective products, and did not reach the realization of the implications of the 1:10:100 rule as quickly as TQM did.

In spite of these limitations, QA, when supported and sustained, can add real value to an organization. These days, much of what the originators of QA developed is now included in quality planning and internal auditing. Also, teams have been consolidated and terms confused, so that I know QA groups that do QC and QC groups that do QA. Nonetheless, we should be sure to include processes of communication, assurance that we are using best practices, and the focus on quality improvement that QA brought to the world of business.

Delivering Quality: Customer Delight

In Chapter 4: *Quality for the Customer,* we defined customer delight as the goal of having every customer fully satisfied with the product and enjoying every encounter with representatives of the company—the elimination of hassle along with the satisfaction of wants, needs and expectations. There, we focused mostly on the process of finding out what the customer wanted. Now, we will look at how we can make sure that they get it.

Customer delight was a late extension of TQM. TQM first worked within the company, then went back along the supply chain to ensure high quality raw materials and inputs. At that point, Toyota made a shocking discovery. Sales were poor in the United States. They did a customer satisfaction survey and found that, although 90% of new Toyota owners said this car was the best they'd ever owned, only 10% said that they would buy another Toyota. The reason was poor customer service. They loved the car, but they couldn't stand the dealers who promised add-on packages and didn't deliver, or promised the car would be ready by a certain date and it wasn't, or who served cold coffee. The distributor for Toyota was an independent company. But Toyota was used to that. They did the same for the U.S. distributor that they had done for their suppliers—provided complete support for a TQM approach. At first, TQM went from the purchased parts to the end of the assembly line. Adding vendor support moved the starting point to the raw materials, Adding customer delight through the sales force pushed total quality forward from the end of the assembly line all the way to the smile on the customer's face. The result: U.S. Toyota customers came to love the cars and the dealers, bringing in a lot of repeat business and referrals.

With customer delight, quality management pays attention to the customer's experience of quality as well as to objective measures of quality. And we do it with integrity, by ensuring genuine value, rather than by the smoke and mirrors that are sometimes used in marketing, advertising, and branding to deliver the perception of quality without the substance.

Key areas of improvement in customer delight include:

- *Training of salespeople, customer service representatives, and repair service people in customer service,* including listening skills, empathy, and follow through.
- *Empowerment of customer service representatives* within the company so that they can do what is needed to delight the customer.
- *Training in PDCA* so that customer service personnel can develop procedures and improve continuously.
- *Team customer service* to increase effectiveness and efficiency while reducing errors.
- *Automated information systems* for customer service, such as customer relations management (CRM) software systems.
- *Automated customer service* that embodies the highest quality customer service into artificial intelligence such as web interfaces and call management systems
- *Intelligent communications and training systems* that provide customer service representatives with the latest solutions and create cost-effective methods for communication with customers, such as support via Internet chat dialog.

A focus on customer delight must be balanced by a focus on the bottom line. We need to make sure our customer delight initiatives are affordable. We also want to be sure that we plan for the benefit of the bottom line by defining measurable goals, such as repeat business and referrals, that are clear measures of benefit to the business. We want to delight the customer—but not by giving away the store.

Conclusion: Quality from Beginning to End

If we succeed in all five stages of quality management, we have quality all the way from our suppliers to the smile on the customer's face. The smooth, high-speed flow of products, services, and information will create a smooth, high-speed flow of cash. Process improvement will result in efficiency as well as effectiveness, so we reduce costs as well. If we aren't getting these kinds of results, then we need to ask why. Investigate the problem, then plan, do, check, and act. When you do, you begin the cycle of continuous improvement. Instead of waiting for the crisis, why not now?

In this chapter, we have covered quality management from one end to the other. In the next two chapters, we will look at the parallel paths of quality—the human side and the technical side. Both are essential to our success.

Q-Ball Quiz for Chapter 6

1. Which of these is *not* a reason that companies should maintain quality management initiatives?
   (a) When we don't spiral up towards more quality, we quickly spiral down towards less.
   (b) If our competitors keep the focus on quality and we don't, we will quickly fall behind and lose market share.
   (c) Quality management initiatives always show rapid results.
   (d) A steady commitment to quality solves other problems.

2. When describing flows in business, which one of these is *not* usually described as flowing? (This is a tricky one. If you get stuck, see the hint after the quiz.)
   (a) quality
   (b) cash
   (c) work
   (d) information

3. SIPOC stands for
   (a) Standard Industry Procedures Optimize Consumption.
   (b) Savvy Individuals Please Opinionated Customers.
   (c) Slick Industry Powerhouses Outwit Consumers.
   (d) Slipshod Irritating Products Outrage Customers.
   (e) Supplier, Input, Process, Output, Customer.

4. Which of these is *not* one of the qualities of a good requirements specification according to IEEE standard 830-1993?
   (a) consistent
   (b) integrated
   (c) feasible
   (d) unambiguous

5. Which of the following statements best describes the relationship between inspection and quality control?
   (a) QC replaced inspection.
   (b) QC provides information for process improvement, where inspection only supports the removal or repair of defects.
   (c) Inspection and quality control are very similar, but QC uses statistical sampling.
   (d) QC replaced inspection because you can't achieve quality by inspecting for defects, you have to control processes to prevent defects.

   Hint for question #2: Add the word "flow" after each answer. Which one creates a term you've never heard before?

Leading a Quality Team

[T]he first object of any good system must be that of developing first-class men; and under systematic management the best man rises to the top more certainly and more rapidly than ever before.
—Frederick Winslow Taylor, *Scientific Management*, 1911.

People make quality. Quality results are achieved only by individuals working both separately and together. In this chapter, you will learn how to help each person on your team focus on quality, and how to pull the team together to deliver quality results.

Leading Your Team to Quality

A quality team is a good team of good people. Upgrading the group of people you work with now to a quality team will almost certainly require change. If you already have a quality team, then change is still essential, because quality teams either follow the path of continuous improvement, or they stagnate. In fact, one of the saddest and most difficult business situations I've seen was in a company dedicated to TQM. But after TQM was achieved, senior management slipped on the education effort and in leadership. The result was a bunch of people surrounded by slogans, thinking that they were in a TQM company, feeling that TQM was a disaster or a joke, and not knowing how to work together to achieve quality.

A team is shaped by its leader. If you're going to lead a quality team, you're going to have to get yourself into shape. Learning leadership skills—self-awareness, self-control, listening, empathy, and the ability to provide guidance—are essential steps for you. As you learn these skills, you can model them for your team, and the team can learn and grow together.

Looking at it another way, a leader is simply a person who has followers. But to lead them in the right direction, we, as leaders, must be going in the right direction. If the right direction is greater ability to define and deliver quality, then we must cultivate these abilities within ourselves, and cultivate our ability to communicate them to others. When we do this, we become able to improve the quality of our company's most important asset—its people.

At the same time, followers make a leader. People willing and able to do good work, and also willing to learn, take direction, and grow make a team—and its leader—successful. We need to find such people and, even more importantly, help each person become a highly professional worker and team member.

What do I mean by a good team of good people?

- Each person is good at what he or she does, and capable of getting better.
- Each person is effective at communicating—listening and expression—about the job and about quality.
- The team leader is an effective self-manager who can grow and wants to grow.
- The team leader has an understanding of the surrounding environment and can offer a vision to the team or help the team create a vision, and then lead the team in the right direction.
- The team has effective methods of meeting and communicating to define status, do planning, brainstorm problems, and clear the air.
- The team has the skills it needs to do the job, or can close gaps by learning and hiring.

Is it really possible to take the team we have, and turn it into a quality team? The answer is yes, sometimes. It is possible, but it is not easy. And not everyone is up to the task. The underlying question is: Can people change? See the sidebar *Can People Change at Work?* for views on this issue.

EYE ON THE BALL

Can People Change at Work?

Almost all management classes present a paradox—a problem of contradiction—and don't resolve it. We hear that to succeed as managers, we need to become leaders, and that requires transforming our character. Then we learn that psychologists agree that basic character is established by the age of eight, and we can't change it. What's going on?

The answer is that what psychologists mean by unchanging character is very different from the qualities of character that we need to have for leadership, which we can learn and develop. For example, say someone is impatient. That is a part of psychological character that is very hard to change or unchangeable. But impatience can appear as angry outbursts, or as a controlled drive towards success that helps others be motivated and focused. The difference is training and practice in self-awareness, self-acceptance, acceptance of others, and effective communications, which are all learnable skills.

The Lesson: We can learn and practice leadership. The key skills are difficult to master, but learning and habit change are possible. One part of that is becoming aware of our character—our strengths and weaknesses—and knowing what we can change, and what we can't change but might instead reshape through self-training.

We can learn the qualities of leadership so that we can lead our team on the path of becoming a quality team. Here are some best practices:

- *Start where we are.* Self-acceptance and acceptance of others, without blame, is essential.
- *Set clear, attainable goals.* Vague slogans and extreme commitments don't work. Setting a clear goal and taking at least one small step towards it every day works.
- *Define and practice your values.* Human memory and focus are valuable and trainable. I have written a personal vision, mission, and values statement, and I try to read it to myself every day. When I do, it works. I see myself becoming more patient, clearer, and more productive.
- *In managing, use a coaching style.* The coaching style of management is most compatible with Theory Y—where we believe that your team members want to do good work—and support them.
- *Gain power from genuine qualities.* Genuine presence and listening, caring about your team while also caring about good work, gains you respect and is positively infectious. We can build on this through effective reward systems of small rewards for clearly defined goals. Referential power can be

gained by knowing that quality management works and guiding the team to learn what you know.

- *Find best practice resources.* In the *Resources for Learning* section at the end of this book, you will find books about leadership, coaching, and professional development. Most of them are either adaptations of TQM to self-improvement. Others apply best practices such as Stephen Covey's *Seven Habits of Highly Effective People* and Emotional Intelligence for Business.

When we are willing to be the first to change, we can rightly ask others to change along with us. In creating quality teams, we see how many people who, though they seem at first to be utterly unable or unwilling to change, become able and willing to change when given clear direction and a supportive environment. Joining together to become a good team of good people is one of the most rewarding experiences in professional life. It is also darned good for our careers and for the bottom line. Become known as a problem solver leading a team that gets the job done, and see where you go from there!

Here are the crucial elements in developing a quality team:

- Quality and Job Definition
- Focus on Quality
- Coaching Individuals to Excel
- Creating a Quality Team

Let's look at each of these in turn.

Quality and Job Definition

The bottom line is people can't do a good job if they don't know what they're supposed to be doing. W. Edwards Deming, a founder of TQM, put it this way, "It is totally impossible for anybody or for any group to perform outside a stable system—Management's job, as we have seen, is to try to stabilize systems. An unstable system is a bad mark against management." (Walton, *The Deming Management Method*)

Are you giving your team a stable system? In a quality team environment:

- Each team member knows what his or her job is, and isn't.

- We all know who prepares our inputs for each task, who we deliver our outputs to, and what their requirements are.
- We all know how our job affects the customer and the bottom line.
- We know what work we will be doing each day and each week, unless something unexpected happens. Unexpected events do happen, but, when they do, we see that we genuinely couldn't have known about them sooner, and couldn't have prevented them.
- We have effective, structured meetings—weekly or more often—to check status, plan future work, clear the air, and make suggestions for how to work smarter.

As team manager, it is your job to put all of this in place with your team. You need to make sure it happens, and they need as much empowerment in defining their own jobs as they are ready for, but not so much that they get lost at sea. See the sidebar *Building a Quality Team* for tips.

Q-PRO

Building a Quality Team
Here are crucial things to keep in mind when building a quality team:

- *Define roles and workflow.* Map workflow using the IPO method from Chapter 6. Define roles as groups of tasks to be done by one person.
- *As much as possible, let the team split up the work the way they want to.* Once all the tasks are defined in the workflow, meet with the team to decide who will do each job. As long as each task gets done, and each person commits to doing his tasks well, that is enough for you.
- *Let each team member define his or her inputs, and their quality requirements.* Then transfer these requirements to be the outputs of the preceding tasks.
- *Let each team member decide how to do his or her own work.* This empowers workers and eliminates micromanagement.
- *Turn the roles into job descriptions.* Always keep up-to-date job descriptions on file.
- *Fill the gaps.* If a task can't be done by the current team, then either provide training or retain new staff or outside services.
- *Define improvement goals with each team member.* Provide small, frequent incentives or rewards for improvements that link to customer delight and improvement in the company's bottom line.

- Build teams to set standards and solve problems. Additional jobs such as discovering relevant external standards, benchmarks, and best practices, or solving specific technical problems, should be assigned to teams—perhaps just teams of two—who use PDCA to come up with a solution by a specific, reasonable date. In a small company or department, it is fine if people are on several part-time improvement teams. Early on, you can guide the teams. Later, your improvement teams will become self-managing.

Lead your team in building a defined, stable work environment that fosters elimination of more and more errors as time goes on.

A full description of how to build a team and define *Standard Operating Procedures (SOP)* is too large to fit in *Quality Management Demystified,* but is available in my other books. See the *Resources for Learning* appendix for further details.

Focus on Quality

We don't usually think of attention or awareness as a business skill. But how often do we want to say to a team member, "Pay attention!" Attention—attention to work, and also attention to our moods and reactions—is a skill we can learn and practice. Creating a team that believes they can improve requires cultivating this skill in ourselves and in everyone on our team.

How do we cultivate the skill?

The first step is to make an environment free of personal criticism, so that people feel safe paying attention, expressing that they care about work, and suggesting improvements.

The next step is to help the whole team understand the benefits of quality work for everyone. We've already discussed the benefits for the customer and the bottom line. What are the benefits for the employees? Quality Guru Philip B. Crosby talks about creating hassle-free businesses. If we define requirements clearly, then we create a situation where, as much as possible, people know exactly what they are supposed to do and when they are supposed to do it.

Everyone receives high-quality inputs, does their job well, and passes on their outputs, contributing to success. And we know that there are no time-wasting dead-end activities that don't deliver value.

I'm not painting a picture of a perfect world; I'm describing a well-managed business that cannot only execute plans but can also handle contingencies. There will be changes to specifications. There will be urgent requests. How many? That depends on the nature of your business, particularly on what is not under your control. If you work in a high-pressure industry where your customers want you to respond to urgent requests, there will be lots of stuff flying at your team. If you work in a stable industry or on a long project, there will be less. What will be nearly eliminated is the unnecessary hassles, interruptions, and changes. You and your team can do this by first setting up a stable environment, then using root cause analysis and permanent preventative actions to improve problem areas.

COACHING INDIVIDUALS TO EXCEL

In delivering quality, nothing can replace each person's attention to work and desire to do good work. It also helps if individuals are more self-managed, taking responsibility for completing work on time to specification. Although specifications are precise, the way we work with each individual must be unique. The coaching approach to management—proven to be the most effective—involves sensitivity to individual differences. And some of the strongest differences come up around how to talk to someone about a mistake he or she has made. Some people handle news of a mistake in a matter-of-fact way and take care of it; others need a lot of support and care. It is your job to do it the right way with each person.

Be proactive. Set up a good working relationship with each team member before you need to work out problems. Help the person get excited about continuous improvement, and then make it specific, defining clearly and specifically where they are now, and agreeing on what improvements they can make in a specific period of time—one or three or six months—that are beneficial to the company. In this context, it is easier to bring up specific errors and even repeating problems. For some people, we need to ask how the person feels; for others, we should avoid this, and let them volunteer their feelings only if they want to. But we should always be specific and clear about the problem and the importance of a solution by a specific date. Vagueness doesn't help anyone solve problems, no matter what their personality. Realistic optimism has been shown to be the most effective for motivating people.

CREATING A QUALITY TEAM

As we discussed in *Chapter 3: Defining Quality,* success of the quality team depends on dealing with ego effectively. The basic framework—that people make mistakes, so errors are inevitable, and we need to work on that individually and also accept team support—was laid out in Chapter 3. Creating a culture that talks about and deals with error directly is essential to success. Here are some next steps:

- *Have the team agree: Don't criticize anyone for being human.* If we all make errors, why should anyone receive personal criticism for doing what we all do sometimes?
- *We each strive to do good work, and that includes talking about and preventing error.* This creates a constructive team environment for quality management.
- *Better the error is found by the team than by the customer.* We set our ego aside to let the team check, review, inspect, and test.
- *Best if we can fix our own errors.* When feasible, the team shows the error to the original worker, who fixes it.
- *Prevent future errors: Fix the process as a team.* The team can see each error as a chance for continuous improvement through PDCA.

QUICK QUALITY TIPS

Focus on the Facts
Fact-based decision making, a TQM tool, is simply an agreement to let empirical facts guide our decisions and discussions. It is also a powerful practical tool to help a team that has lost focus and fallen into blame or confusion, especially if the team has decided to use this approach in advance. To clear the air, you can say, "We all have strong feelings about this. But can we clear the air quickly and let the facts speak for themselves?"

When a team is on the path of continuous improvement together, synergy arises. Synergy is the quality of people being able to do more together than we could do separately. A hassle-free work environment makes room for synergy. One thing that helps synergy appear—it always appears as a gift, we can't force it—is customer service on the team. When we realize that our outputs are inputs for someone else, and we strive to do good work so that the next person can focus on his or her job, instead of running into a hassle getting what he or she needs, that is customer service on the team. With a well-designed workflow and customer service on the team, each person sees how he or she is a valuable part of the team, and how each task adds value for the customer or the company. Through the workflow, everyone is connected to the customer. In this environment, people can take genuine pride in their work and appreciate support from the team—even when that support means hearing about their mistakes.

The sound of a quality team at work is often silent focus punctuated by expressions of joy.

Conclusion: The Quality Team and the Soft Side of Quality

To those who want to reduce quality management entirely to a science, applying statistics and achieving ever higher and higher levels of Sigma, I quote Shakespeare: "There are more things in heaven and earth, Horatio, than are dreamt of in your philosophy." Just as Smeaton—for all his good engineering—made a stronger lighthouse by imitating the shape of an oak tree, without knowing why it would work, so we must realize that each person's contribution to the quality team—definable and measurable, or not—is essential to success.

And I would also say, "You're right. There is a lot of value in good technical work and precise engineering." To learn about the engineering side of quality, let's move on to *Chapter 8: Quality Engineering*.

Q-Ball Quiz for Chapter 7

1. Which one of these is an effective part of leading a quality team?
   (a) Slogans.
   (b) Making sure everyone takes care of all the details.
   (c) Improving the team's ability to pay attention.
   (d) Making sure you treat everyone exactly the same.

2. Which is the best statement about individual and team quality effort?
   (a) Individuals do their best, and the team provides support.
   (b) As individuals, we make mistakes; so we need a team to do quality.
   (c) Both individual attention to quality and a team effort to catch errors are essential to quality work.
   (d) We need teams because people can't manage the quality of their own work.

3. Which of these quality management methods has been adapted to be a technique for personal transformation in quality work?
   (a) Total Quality Management
   (b) Six Sigma
   (c) Scientific Management
   (d) The Zero Defect movement

4. When telling a team member about an error, which of these should you *always* do?
   (a) Ask the person how he or she feels.
   (b) Be clear and specific about the facts.
   (c) Give the person the chance to fix the mistake himself or herself.
   (d) Tell the person in private, without the team present.

5. Which of these is *not* true about synergy?
   (a) It arises, but we can't make it happen.
   (b) We can create the situation where synergy arises by eliminating hassle and focusing on quality as a team.
   (c) The attitude of customer service within the team makes synergy more likely.
   (d) As a team, we can make synergy happen.

Quality Engineering

Science, not rule of thumb.
—Frederick Winslow Taylor, *Scientific Management*, 1911.

Now that we have a good requirements specification and a team dedicated to quality, we have created the environment in which quality engineering can really add value to our customers and to our company. Quality engineering is essential. But, alone, it is not enough. When our team does quality engineering, and the quality management process connects the technical or engineering team to the customer through good requirements management, quality engineering pays off.

EYE ON THE BALL

Putting Together the Pieces of Quality Engineering
Quality engineering includes making quality definable and measurable and ensuring we apply quality methods from product design all the way through development to delivery. We can also focus on continuous improvement with our quality team and reduce the cost of quality products and service through automation and robotics.

Definable Quality

Once we have a complete requirements specification—including requirements from customers, stakeholders, standards, and regulations—we face the difficult task of ensuring that each requirement is defined well enough to be objective. A requirement is *objective*. It is so well defined that anyone reading the requirement and examining the results of checking would agree whether or not the item meets the requirements. The opposite of objective is subjective, and something is *subjective* if a change of the person performing the test or work of checking would change the result.

In introductions to quality engineering and statistics for quality, objective is often confused with measurable. Consider this example. Say we produce polo shirts in three colors—black, red, and blue. Before an order is shipped, one quality check is to make sure that the color of the shirt is correct. Now, anyone looking at the shirt—worker, supervisor, or customer—will agree if it is black, red, or blue. In this case, examination is necessary for objectivity, but measurement is not.

When the attribute under observation is an amount we can determine by measuring a continuous variable, then measurement is necessary. We must add tolerances to the requirement, and then the requirement is likely to be objective.

Oddly enough, we can use color as an example of continuous measurement, as well. Say we produce video cameras. One test might be to aim the camera at a color palette under controlled lighting, then use an optical testing device to determine that each pixel (dot on the camera's video screen) is in the right range of brightness, shade, and hue. In this situation, color is a continuous variable measured by machine. I chose these two examples—both using color, but only one requiring measurement—to point out that our practical needs for the requirement determine how we define it and how we make it objective. How to observe or measure an attribute is an engineering decision involving the nature of the object and the work to be done, not something intrinsic to the nature of the product.

Previously, I said that a measurable requirement defined with tolerances is likely to be objective. Why only likely? What are the exceptions? Well, suppose we have two different measuring devices that use a different technique or are calibrated differently. Then they might produce different results. This is not a small issue—it is part of the reason why the Hubble Space Telescope went into orbit with a bent mirror, and that was a very expensive defect. Quality engineers—with reference to standards—must determine how checking will be done, and ensure that the method of checking is accurate and unbiased enough to ensure that our measurements are objective. A measurement is sufficiently *accurate* if it can measure closely enough to ensure that the attribute is within tolerances. For example, if our tolerances are within a tenth of an inch—say the requirement is "between 3.90" and 4.10" inches—then we should measure to the nearest 100th of an inch. Note that the number of digits we write in our specification—called significant digits—can be used to indicate the required degree of accuracy. A measurement is *unbiased* if it is free of an error that consistently throws off the results in one direction or the other. For example, if we need to measure weight to the hundredth of a gram, then dust on the scale would add significant weight, biasing the results so that the numbers we saw were always higher than the actual weight of the item.

This discussion just begins to scratch the surface of what quality engineers have to address when defining the best way to perform a single check of one product against a single requirement. Now imagine that we are producing thousands of cars a day, each with tens of thousands of parts, each of which has to meet many requirements. Quality engineering is the discipline that puts all of that together and makes it work effectively, efficiently, and affordably, beginning to end.

End-to-End Quality

In Chapter 5: *Key Quality Concepts,* we explained how, because of the 1:10:100 rule, each error costs less the earlier it is removed from the product during its lifecycle. We can carry this even further by prevention, which is the act of removing potential errors. As we discuss quality from the beginning to the end of the production process, keep in mind that we get a bigger bang for the buck by focusing more effort on the earlier stages.

DESIGNING IN QUALITY

We often hear the phrase *designing in quality,* but what does it really mean? It might be easier to think of designing out opportunities for error and designing out potential defects. This will mean different things in different situations. For example, if seams between parts leak, then we can design in quality by redesigning our product to have fewer parts. When we do that, though, it makes each part larger and heavier. If large, heavy parts break more often, then we can design in quality by redesigning our products to have fewer parts, each one lighter in weight. But then, of course, we have more seams. As you can see, designing in quality requires knowing defects and their sources, and then eliminating the sources of defects.

Processes for Designing in Quality

Over the years, quality management engineers have developed tools that help us to eliminate possible error and design in quality. On this list, items in **bold** are statistical methods discussed in Chapter 10.

- *Track all defects.* We can't redesign to eliminate errors until we know what the errors are. We need to track all defects found in review, inspection, and testing. We also need to gather information about products that have already been sold to customers, to see how they are holding up.
- *Calculate the cost of the consequences of each defect.* We need to determine which defects are most costly considering consequences of the defect if it reaches the customer, difficulty and cost of catching and correcting the defect, and frequency of the defect.
- *Find multiple causes for single defects* with **Ishikawa—also called cause and effect, or fishbone—diagrams.**
- *Prioritize defects for the prevention effort* using **Pareto optimization.**
- *Brainstorm possible solutions.* See the example in the next section.
- *Evaluate the benefits and costs of each solution.* We forecast: How effectively will the new solution reduce or eliminate the defect? How much will it cost to implement? How will it change the cost of production? What new potential errors will it introduce? This cost/benefit analysis will allow us to choose one option.
- *Plan, develop, and implement the design change.*
- *Trace our results.* We can't just assume we've solved the problem and introduced no new problems. The continuous improvement cycle of PDCA never stops.

Q-UP

Evaluate Your Quality Engineering
Here are some questions to ask to decide the next step in improving your team's quality engineering, with directions toward solutions.

- *Do you have enough communication across departments to design in quality?* If not, your company may need a quality improvement program using one of the methods from Part 3.
- *Are you able to form effective quality improvement teams?* If not, see Chapter 7: *The Quality Team.*

- *Do you have a structured method for defining sources of error and eliminating them?* If not, use the bulleted list on page 118, a method from Part 3, and standards and methods from your technical or engineering discipline to create one with your team.
- *Are you missing any of the steps from the bulleted list of steps and tools?* If you are, read the appropriate chapter to learn about the tool, then create a quality improvement team to investigate best practices and implement a solution.

Finding Design Solutions

There are many possible solutions to almost every design problem. As an example, let's pick up with the device we discussed above. It might be something like a washing machine or an air conditioner. (For an excellent book on quality management and quality engineering that uses a case study from the air conditioning industry, see David A. Garvin's *Managing Quality: The Strategic and Competitive Edge* [1988]).

Here's the situation: The air conditioner we sell can leak at soldered seams, but, if individual parts are too heavy, they can crack. So, we have both problems. Here are a bunch of design solutions.

- Using current technology, combine parts to make fewer parts, up to the point where the cost of cracks is greater than the cost of failed seams.
- Determine when parts crack. If they crack on the shop floor, rearrange the shop floor and improve handling to reduce cracks. If they occur in shipping, improve internal support for parts inside the unit and improve packaging. Once these changes handle the problems with cracks, we can make larger, heavier parts, and solve the problem with failed seams.
- Experiment with new materials or new structural design to reduce parts cracking.
- Experiment with new methods for making seams that fail less often.
- Look into entirely new materials and technologies, such as materials molding, and prototype a solution that has lighter, less vulnerable parts, fewer seams, or maybe even self-repairing seams.

With current technology, we have only the first option, which is a trade-off between the two problems. But if we are willing to re-engineer our process or product, we can reduce both problems at once. Those kinds of solutions require an in-depth, up-to-date knowledge of the technology and issues of our industry and outside-the-box thinking. That's what we mean by designing quality into a product or service.

Now, let's return to a situation where we have a defined product and a defined method of production. In a stable situation, how do we manage and improve quality?

BUILDING THE QUALITY PLAN

Quality engineering efforts should be organized. The first step is to build the requirements matrix, which links original requirements from any source to the attributes of components that must be of quality to deliver to that requirement. A single customer requirement may involve many specific technical requirements for different attributes of our product and different steps of our process. For example, let's say that we manufacture ham-and-cheese sandwiches in large quantities to be delivered to convenience stores. Let's say we have a quality requirement that our sandwiches not make any customers sick. That one very reasonable customer requirement translates into the following technical requirements:

- Ensure incoming ham is fresh and has always been stored at acceptable temperatures.
- Ensure incoming cheese is fresh and has always been stored at acceptable temperatures.
- Use a bread with appropriate preservatives.
- Make sure our facility is clean and health procedures are followed.
- Make sure that our process doesn't run into delays, and that ham and cheese are always stored at cold temperatures.
- Wrap sandwiches completely and tightly.
- Make sure refrigeration on all trucks works, providing a stable temperature for the sandwiches.
- Make sure that there are no significant delays when loading and unloading sandwiches.
- Ensure that condiments and other ingredients are chosen with an eye to minimizing common allergens.
- Ensure that product labeling warns about any significant allergens.
- Ensure that stores use appropriate refrigeration and maintain proper temperatures.
- Ensure that stores discard sandwiches by the expiration date.

Those are eleven separate technical requirements all needed to meet one customer requirement. This is an example where the sandwich-making process is a system, and any of several different failures in different parts of the system can cause the undesirable outcome. Can you think of any more requirements to keep our sandwiches safe?

TRANSLATING REQUIREMENTS INTO TESTS

Maintenance of each of these requirements needs to be checked. This will require review, quality control with inspection and testing, and quality assurance.

- We would review SOPs to make sure that they met health requirements, and purchase orders to make sure they specified product requirements to the companies that supplied ingredients.
- Quality control could be achieved by taking temperature measurements and comparing them to requirements—a form of inspection—and by sampling sandwiches and testing for growth of bacteria or mold.
- Quality assurance would involve audits across the entire production and delivery chain. We could check to make sure truck refrigerators were working and that stores were throwing away sandwiches on time.

Some tests may be mandated by regulation. But the entire plan for checking and testing needs to be designed by quality engineers. The goal is to perform all of the necessary tests to bring error down to an acceptable level—in the case of food poisoning, we aim for zero defects—at the lowest possible cost.

TESTING BEFORE AND AFTER DELIVERY

Although we can reduce cost of quality by focusing our efforts on prevention and early detection, we should not ignore the fact that the last chance to prevent errors from reaching the customer is just before delivery. As a result, testing after final assembly, QC and QA for delivery procedures, and, where appropriate, testing at the customer site, are all very important. Although this is obvious, it needs constant reinforcement. Workers involved in final assembly and delivery are often under pressure to focus on productivity, speed, and meeting a customer's immediate demands. All of these can distract from a necessary focus on quality. Also, if products go through middlemen—whether delivery companies or stores—then we have less control over the quality of work of the people who don't work for us. Consider the following scenarios and solutions that apply quality engineering to the product during and after delivery.

- A company that produces computers might select a single vendor for shipping, and then require that vendor to maintain a very low rate of damage in shipping, or risk losing the contract.
- A small company came up with a great system: one penny was added to a jar for every order shipped, but a dollar was taken out for every customer complaint of an error in packing and order fulfillment. The shippers got the money in the jar as a bonus every month.
- Instructions for unpacking and assembly should be carefully designed and tested so that they work for the customer, reduce the chances of customers damaging equipment, and provide immediate solutions if the product was damaged in shipping.
- Computer equipment now often comes with diagnostic software that can report problems to the manufacturer over the Internet.

A combination of effective technical instructions and human communications catches error before it is too late, and provides support when the customer does find a defect.

Leading Quality Engineering Efforts

Everything we said about quality teams is true of teams in technical and engineering environments. There are some plusses and minuses worth noting.

- *Continuous improvement with PDCA is often easier* because engineers are more likely to have encountered the scientific method or statistics in school, or quality management methods on the job.
- *Connecting to the customer is often more difficult* because the distance between the truly technical engineer—who knows an awful lot about a specialty, measurement, and data analysis—and a customer is a lot farther than say, the distance between a retail salesperson and the customer. There is less contact, and the technical jargon of the field can get in the way.
- *Corporate structure may be a barrier to quality improvement efforts,* making it difficult for engineers to get the information they need from field representatives or suppliers. It may also make it difficult for engineers to address issues or provide solutions to people outside their own work group or division.

- *Within a department, quality control or quality assurance can be isolated.* QA and QC groups are often demoted to the job of checking other people's work, and then resented for showing up other people's mistakes.
- *Some people go into engineering because they want to work alone,* in which case, building a true team environment is difficult.

Managers and executives need to identify and solve these problems. These problems, which are still with us, were identified as weaknesses of North American quality assurance programs in the 1980s in contrast to TQM. TQM prevented these problems by making a top-down, company-wide process of improving quality methods. Executives led managers and engineers, emphasizing teamwork and a focus on quality for the customer. In fact, in TQM, a good quality improvement program moves beyond the limits of the company. Employees are empowered to negotiate with, and even train, vendors on issues of quality. Also, in TQM, every engineer is a quality engineer, and every employee is a quality team member. Specialists in quality are not isolated. They are teachers respected for their ability to help others do a good job.

Automation, Robotics, and Quality

The earliest uses of automation and robotics were for increased productivity at lower cost, and then to automate processes in workplaces that were uncomfortable or dangerous for people. The invention of powered devices with automatic controls replaced the blacksmith with the steel foundry. Computer-controlled machinery brought in a new level of automation, and, by the 1970s, there were plants that made photographic film running entirely in the dark with no people inside. Computer-controlled machinery coated plastic sheets with layers of light-sensitive chemicals to make film for cameras and X-rays. Quality control was automated into the feedback of information into the computers that controlled the machine process. More recently, I've visited a factory that is just one giant machine. Sand and chemicals are poured in at one end. The sand is melted into glass, formed into shape, and coated with chemicals. Finished television tubes come out the other end. Quality control measurements are built into the machine, which adjusts itself to maintain production within tolerances. The company that owns the factory can focus on issues such as how many of which tubes to produce, rather than the details of ensuring that each TV tube meets quality standards.

General advances in computing and communication have supported quality improvement efforts by reducing communications costs and making it easier to gather, organize, and analyze information with databases, data mining, and advanced statistical analyses.

With the arrival of the Internet, we are seeing automation solving new challenges. How can a web-based clothing store like Land's End compete with stores where you can try on the clothes and see how they fit? Go to their website and check out the virtual model. You measure yourself and enter your specifications. Then you can pick clothes from their catalog and see how they look on you. Setting up the model takes less than half an hour, and then you can try on clothes as often as you like and even email the picture of your virtual self in clothes in your shopping cart to a friend or fashion consultant. It's not the same as trying on clothes in a store. But it fulfills some of the same function, and does it with computer automation replacing the cost of retail space and expertise of sales clerks.

Conclusion: Engineering for Continuous Improvement

Engineers have a central role in quality control and quality assurance. It is up to managers and executives to remove the barriers that can sometimes separate engineers from customers, other teams, and vendors. When we make the connections, engineers can contribute to increasing quality and reduced cost all the way from raw materials to the smile on the customer's face. Companies that have continuous improvement in place spend less on production and meeting standards, so they can invest in developing new solutions to reduce cost, such as computer automation and robotics, and improve quality.

Q-Ball Quiz for Chapter 8

1. Which of these is *not* a barrier to including engineers in organization-wide quality improvement?
   (a) Engineers often have their own jargon and can't speak in ways customers understand because they haven't learned how.
   (b) Companies often have barriers between divisions that isolate engineers from field service data and contact with suppliers.

    (c) Engineers only want to focus on reaching measurable goals and don't care about customers

    (d) Quality Assurance and Quality Control engineers are sometimes put in the role of being testers who just tell other people what they did wrong.

2. An objective test result
    (a) is the result of testing a physical object.
    (b) would be the same no matter who performed the test.
    (c) could be explained to someone who disagreed with it, so that they could not object.
    (d) does not require a person, as it is performed by automated machinery.

3. A good measurement in a quality test is all of these, *except*
    (a) always used to test every part.
    (b) unbiased.
    (c) as low cost as possible.
    (d) sufficiently precise.

4. Which of these is *not* true? Testing just before delivery
    (a) costs too much; the 1:10:100 rule dictates that all tests should be done sooner.
    (b) is important because it is the last chance to catch errors before the customer.
    (c) reduces returns and warranty repair costs.
    (d) can be automated.

5. Which of these techniques would *not* likely to be central to re-engineering to design in quality and eliminate error?
    (a) Quality Assurance methods
    (b) Pareto diagrams
    (c) Root cause analysis
    (d) Quality control testing

CHAPTER 9

Auditing Quality

*[A] careful analysis was made of every element
of this machine in its relation to the work in hand.*
—Frederick Winslow Taylor, *Scientific Management,* 1911.

Auditing is a misunderstood and undervalued profession. When we hear the word audit, we usually think, "tax man." Or we might think of the auditors of Enron or WorldCom who fell down on the job. Few people realize that there is an entire world of internal auditing, and another world of quality auditing, which have nothing to do with taxes and very little to do with financial auditing. The key function of internal auditing, as illustrated in Figure 9-1, is to add value to an organization by increasing its ability to perform its mission, reducing cost, and reducing fiduciary risk. Auditing is not always about standards. Audits can evaluate operations or processes to help define beneficial management or technical practices or even best practices. We can seek—perhaps struggle—to keep processes up to standard. Or we can work to standard and seek to improve the standard incrementally through an approach called *continuous improvement* or *kaizen.* Or, we can seek—through external research or internal innovation—for a breakthrough to a new level of best practices.

When performing process audits, auditors often work cooperatively with managers. Quality auditing seeks to ensure that quality practices are meeting internal and/or external standards. Internal quality auditing would ensure conformance to internal standards, and also would assist with preparation for formal certification. External quality auditing would be involved in the assessment of

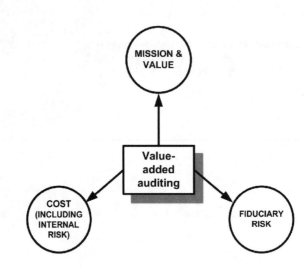

Figure 9-1.   Value-added auditing.

certification, such as ISO-9000 or CMMI, or the granting of awards such the Baldridge Quality Award.

Audit services departments are very common in government and university settings. Any large business would be well advised to research the value of internal auditing if it hasn't already, or to make sure that auditors are not limited to auditing financial functions. Auditing is a crucial part of achieving quality in business. Auditing offers two opportunities for improvement. The first is a chance to assign independent resources where they are needed most, either because an independent perspective is lacking, or because resources are not sufficient to assure quality. The second is the chance to look back at what went wrong and learn what our mistakes were, and where to focus process improvement efforts.

EYE ON THE BALL

Auditors Investigate and Maintain Independence

Independence—the ability to provide an unbiased perspective because they are not involved in decisions and do not benefit from a particular course of action—is essential to effective auditing. As a result, auditors do not manage. Auditors investigate, report, recommend, and sometimes advise. In terms of the PDCA cycle, they perform a checking function, and they may recommend a solution or work with management to define an acceptable solution. But the corrective action is left to management.

Why? Because, if auditors took the corrective action, who could check their work?

Adding Value and Managing Risk

How does an audit department decide what to do? There are always more depart-ments—and more potential problems—to look at than there are hours of audit worktime in a year. The audit department starts with a goal: adding maximum value to the organization using available audit resources. As we see in Figure 9-1, there are three basic ways to do this.

- *Increase the organization's ability to perform its mission.* In business, this means things like selling more products, developing new products, increasing market share, and moving into new markets. Here, auditors pro-vide information to increase an organization's *effectiveness.* This includes reducing the risk—the possibility—of failure.
- *Reduce cost.* An organization is better off if it is more *efficient,* that is, if it can do the same job with fewer resources. Auditors can identify wasted resources and effort. This includes reducing the risk of wastage, as well as actual wastage.
- *Reduce risk of fiduciary liability.* Every organization, in addition to its own risks of failure or wastage, has risks related to others. These are called *fidu-ciary risks* and they are a *liability*—a potential cost—to the organization. For example, every company has personnel files, and most have customer credit card information. If the company allows this information to fall into the wrong hands and be misused, then the company could be sued—it would be liable. Auditors can evaluate these risks and recommend man-agement action to reduce *exposure,* that is, vulnerability to lawsuits, crim-inal charges, loss of reputation, or other costs.

One of the real benefits of the model in Figure 9-1 is that audit departments can sell the value of their services to internal customers—executives and man-agement. The value-added model allows auditors to speak the language of busi-ness—mission, net revenue, and risk—instead of the language of regulations and standards, which is more removed from perceived value.

How does the audit services department decide what to do? Each year, the department creates an annual audit plan, often using a technique called the risk footprint. Auditors perform structured interviews with senior management, look-ing at these issues:

- *Size of the operation.* The amount of money to be earned, or money being spent.

- *Assessment of risk for particular activities.* Essentially, auditors ask managers how well is it going and what are the chances of success? Activities with a greater chance of failure are rated higher in the risk footprint.
- *Other audit or assessment activities.* Sometimes, another audit department or agency is already overseeing or evaluating a function. In that case, the audit services department may assist, but will not duplicate the work.

The result of this series of interviews is a *risk footprint* for the organization, which prioritizes the likelihood of valuable results of audit work in different areas of the company. Evaluating risk is parallel to looking for error—we don't know if anything is wrong until we check. But the risk footprint is the best assessment of which rocks are worth lifting up and peering under. The managers of the audit department then assign resources, planning audits of the highest-risk areas, and looking for ways to do surveys or reviews of as many other areas as possible. One technique that is becoming more popular is to have directors and managers certify that all work—or particularly risky types of work—are being properly managed. For example, if an audit services department doesn't have enough staff to audit every IT development project, then auditors would be assigned to large projects and those known to be in trouble, and senior management would be required to certify that other projects are under control and likely to succeed. Before a manager signs his name to something like that, he's going to take a look.

Some audit resources are available for special investigations at management request, and can also be available to advise on best practices in advance, when a project or management initiative is just getting started, and before it gets into trouble.

Auditing Standards and Methods

Auditors work in compliance with standards created by professional associations and governmental agencies. The standards are detailed and precise, so it is difficult for anyone outside the field to really get a sense of what is required for an auditor to do his or her job. Following the standards assures a very high level of independence. Following standard audit methods allows for in-depth investigation of risks and problems, and discovery of issues that might not be uncovered any other way.

Different Standards

In our society, there are many types of standards. Understanding them is valuable to understanding the costs and challenges of quality management effort. Here are some of the most important types of standards, starting with the strictest:

- *Philosophical.* For an argument to be accepted in philosophy, it must meet an extremely high standard of rigorous logic and definition. That is why some philosophers can continue to say that quality is indefinable even though, on a practical level, we define it successfully.
- *Scientific.* Scientific standards are also extremely rigorous. One view holds that nothing is ever proven in science—not even the law of gravity. Any law or theory is just the best model we have right now. New theories or models might be proven better with new evidence or new experiments. Note that philosophy and science can continue for centuries without being sure that they are right.
- *Forensic.* The original and primary meaning of forensic is that the investigation or approach is acceptable in a court of law. Most forensic work, like quality management, now applies the scientific method. The standards are rigorous, but the time limit imposed by the practical needs of the legal system makes them less rigorous than the standards of scientific research. Auditors must meet forensic standards, and forensic standards are the most demanding standards required of most businesses.
- *Engineering.* Engineering standards have a simple goal—to produce things that work in the physical world. When we don't meet engineering standards, things break down, and we—as a business, with our customers, and in society—pay the price. When engineering standards are wrong, everyone who follows the standard pays the price. As Nobel Prize-winning physicist Richard P. Feynman put it in his "Minority Report to the Space Shuttle *Challenger* Inquiry" in *The Pleasure of Finding Things Out* (1999), "For a successful technology, reality must take precedence over public relations, for nature cannot be fooled."
- *Business.* Standards of business practice by industry and culture. From a quality management perspective, we should meet them routinely and exceed them when the benefit outweighs the cost.

Because auditors work to a forensic standard, their work is slower and more costly. Also, they need more complete and precise information than most business people realize or can fathom. When I teach auditors, I tell them that, when they ask for all the records related to an issue, the auditee doesn't understand the word "all" in the same way as they do. For an auditor, "all" means everything, plus an explanation of anything missing. In business, "all" usually means "everything I can get my hands on easily, and don't worry about the rest." So, when working with auditors, please give them what they need—it reduces hassle and cost, and increases the net value of their work.

The lesson: In quality management, we should be aware of the level of standard we are working to, and communicate clearly about the needs of people working at different levels.

The process of the auditor's own work is always governed by the standards of the audit profession. In addition, auditors may apply standards to the work of the auditee client. In auditing, the auditee, or client, is the department whose work is being examined or investigated. The customer is the recipient of the audit report, usually a board of directors or an audit committee of the board, but sometimes senior management. In some situations, audit reports must be made public. As a result, there can be a lot of politics surrounding the work of auditors. From a quality management perspective, this is unfortunate. Politics is basically noise that gets in the way of understanding the facts; without the noise, we can do better work.

The scope of the audit as defined in the annual audit plan is very broad. It must be narrowed and specified in the early part of the audit project, before the auditors get deep into their work. In most internal audits, this is a cooperative process where the auditor works with managers to define where problems are most likely to be found.

In addition to scope, auditors define clear objectives for the audit. Objectives are sometimes expressed as questions, or sometimes open with a phrase such as "determine whether." When they get down to work, auditors examine controls, processes, or work results, depending on the type of audit. This examination usually falls into one of four categories, based on the criteria stated in the objective:

- *Comparison to an appropriate standard or regulation.* An objective might be: "Determine if processes within the factory that produce our ham-and-cheese sandwiches are in compliance with relevant FDA regulations."
- *Comparison to best practices.* An objective might be, "Define the best practice for ensuring that ham-and-cheese sandwiches are properly marked with expiration dates, and make recommendations to help management adopt best practices."
- *Benchmarking, that is, comparison to accepted targets for effectiveness or efficiency.* An objective might be, "Measure the time it takes each driver to unload ham-and-cheese sandwiches and stock the fridge. Then provide management with guidelines for internal benchmarking of unloading and stocking activities."

- *Technical analysis applying common sense, such as the elimination of work that delivers no defined result and the elimination of duplicate effort.* For example: "Define the process flow of assembling a ham-and-cheese sandwich, and recommend specific changes to increase efficiency."

In doing their work, auditors use many of the same tools found elsewhere in quality management. They make a flow chart or data flow diagram or process diagram of actual work, then compare it to the standard of how the work should be done, perform experiments, or simply analyze work processes and recommend more effective or efficient practices. They use statistical analysis to find correlations and root cause analysis. Two things make audit work different from—and more valuable than—routine quality assurance. First, the greater independence—because the auditor is not part of the team, and only there temporarily—can allow the auditor to see and bring to light assumptions no one else has seen. Second, the auditor generally has more time to do in-depth research than is available to most QA teams.

The results of a classic audit are expressed in a finding that has four parts. Here is an example from performance auditing:

- *Condition:* The measured or assessed level of performance of how things are or were done.
- *Criteria.* What should or could be. For example, the relevant standard or best practice.
- *Effect.* The gap between condition and criteria, and its consequences.
- *Cause.* The reason for the good or poor performance. In some cases, it is not cost effective, or not relevant, to determine cause. For example, if people who were managing and performing sub-standard work have all left, the new management needs only to know how things are being done, how to do them right, and the consequences of not cleaning up the mess.

An audit involving an experiment is one where an intervention is used. That is, the auditor measures the results of a specific practice. Ideally, the experiment has an *experimental control,* where the same process is performed with as close as we can to having just one change—the intervention. The finding of this type of audit is expressed in an *Impact Finding,* as follows:

- *Condition with the intervention,* that is, the results of the new process.
- *Condition without the intervention,* the way things were done in the experimental control, or before the experiment.

- *Effect/impact.* The results of the intervention, that is, the difference in result between the intervention and the control.
- *Cause.* The intervention—the new process or program—is the cause.

The finding or findings are the core of an audit report. However, it is a mistake to think it is good or bad if an audit produces findings. For example, if an audit checks to see if we are meeting regulations, and we are, then there are no findings, and that is a good thing. If an audit tests a new procedure with an intervention, and finds that it makes things better, then that's a finding, and that's a good thing. What we want from an audit is not findings. We want clear, factual research results that we can use to improve our business.

In some cases, especially in internal audits, the audit report also makes recommendations. Another approach is that, once findings are clear, the auditors meet with management, and management prepares a *management response* that includes auditor recommendations into a plan of action—usually with a commitment to fix the problem or achieve certain results by a certain date. Working this way, management integrates the work of the auditors into its PDCA cycle.

Auditing to Quality Standards

There are a number of circumstances where auditors evaluate our process or results to see if they conform to quality standards. Here are the three most common reasons:

- *A company commits to a standard.* When a company or division commits to following a certain standard, then there will be a number of internal and external evaluations. Internal audit can assist by performing checks and tests to let management know if it is ready for certification, and how it is not. The external certification process will either grant certification, recommend improvements before retest, or deny certification.
- *A client or other stakeholder requires a standard.* By contract or, in the case of government clients, by regulation, a client may be authorized to perform an audit or have an outside agency perform an audit. For example, the U.S. military is now requiring major contractors and most subcontractors who provide software to the military to have CMMI certification, as discussed in Chapter 14: *The Capability for Quality: CMM and CMMI.*

- *Our products or services may be operating in an audited environment.* For example, we may offer software and an Internet-based update service that is not usually audited. But a client—perhaps a government agency or major corporation—might perform an IT security audit that reveals findings that our software or services, or their current implementation, represent a security risk for their organization. Audit standards give auditors for a client authority to request information and cooperation from vendors.

In the first situation, it makes sense for the organization to develop its own internal audit capability in relation to the standard the company seeks to maintain. It is highly cost-effective for a company to be able to educate all of management and appropriate workers in the standard and then build effective communications between management and internal auditors. Using this approach, a well-managed quality improvement program that includes achieving certification can meet organizational goals and pass certification with flying colors.

Q-PRO

Managing to Cooperate with Auditors

Executives and managers have a great deal to gain by cooperating with auditors and integrating internal auditing with operations management. There are four ways to do this:

- *Teach management to make good use of audit services early on.* For example, project managers can contact audit services early on and ask: On this project, and with this product, what standards must we meet, and which best practices would be valuable? Also, what controls should we have in place and what information should we maintain that will make a later audit simple for you? Because of the 1:10:100 rule, four hours of audit services time answering these questions could eliminate the need for a 400-hour audit later.
- *At the beginning of an audit, the highest-ranking executive should direct full communication and cooperation.* Managers and workers need to know that cooperating with the auditors makes their own boss happy. This gets information flowing faster, reducing audit costs and hassles.
- *Create a structure that uses audit information to achieve management objectives.* Because they must maintain their independence, auditors can't get heavily involved in consulting or support for actual improvements. So it is useful to have another group, such as a Project Management Office (PMO), a quality assurance group, or a cross-functional team assigned to review audit results and support or ensure implementation of audit recommendations.

- *Encourage auditors to provide "leave-behinds."* Auditors can leave behind data gathering and reporting programs, measurement tools, and evaluation methods that management can incorporate into ongoing operations. This makes it easy for management to bring work up to the standard the company wants to maintain.

Conclusion: Adding Value Through Auditing

Auditing is an opportunity for business to take advantage of experts' in-depth research. Using audit services, executives can increase effectiveness, increase the ability to achieve defined missions, increase efficiency by reducing cost or increasing productivity in relation to critical factors, and reduce fiduciary liability. Establishing a cooperative relationship between management and auditing is challenging due to historical misunderstandings, fear of publicity, and misunderstanding of the audit profession. However, the executive who wants his company to achieve quality will work to overcome these barriers and ensure that the business receives the benefit of value-added auditing and achieves and maintains desired certifications.

Q-Ball Quiz for Chapter 9

1. Which of these is *not* one of the three main ways in which auditing can add value?
   (a) Increasing ability or likelihood of achieving mission.
   (b) Reducing cost or risk of cost.
   (c) Holding management accountable.
   (d) Reducing fiduciary responsibility.

2. Which of these best describes an auditor's activities in terms of the Plan Do Check Act (PDCA) cycle?
   (a) Planning to reduce risk.
   (b) Doing activities that improve management.
   (c) Independent checking.
   (d) Defining action to resolve problems.

3. A *risk footprint* is
    (a) a sign that Sasquatch—also known as Bigfoot—is nearby.
    (b) the mark left on a manager's behind when he is kicked out for risking too many corporate assets.
    (c) an auditor's assessment of the likelihood of the company being sued.
    (d) a tool prepared by auditors using structured interviews to determine what audits or reviews are most likely to have significant findings.

4. *Forensic* means
    (a) related to, or at a level acceptable for, legal proceedings.
    (b) scientific.
    (c) an autopsy or post-mortem—investigation after the fact.
    (d) of or related to standards.

5. An *intervention* is
    (a) an external audit required due to suspicious activity.
    (b) a new or different process being tested by evaluation or experiment.
    (c) a tool for resolving disputes between auditors and managers.
    (d) one part of a traditional finding.

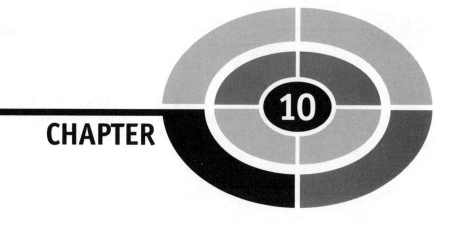

CHAPTER 10

Statistics for Quality

knowledge so collected, analyzed, grouped and classified into
laws and rules that it constitutes a science
—Frederick Winslow Taylor, *Scientific Management*, 1911.

Statistics is valuable to quality management because it gives us a sound basis for making decisions such as:

- Is this batch good enough to meet customer requirements?
- Which of these changes will eliminate the most number of defects?
- What change should we make to our process so future production meets customer requirements?

Statistics is an independent discipline that fits well with the scientific method. It draws from—but is not part of—philosophy, science, and probability. It is applied in science, engineering, psychology, business, and other fields. In the broadest sense, the field of statistics is a structured way of saying things about how the world works. In practice, its two most important benefits are:

- Statistics allows us to view a part of the world—a sample—and conclude things about a larger part of the world—the population—that we can't see and measure affordably.
- Statistics can be used to take information about the past and forecast the future.

Statistical analysis has strict requirements, so we can't always use it. We'll open Chapter 10: *Statistics for Quality* with a discussion of when statistics don't apply. Then we'll look at key statistical concepts. Lastly, we'll look at common statistical tools for quality management.

When Statistics Doesn't Apply

Historically, in understanding statistics for quality management, there has been a lot of confusion around terminology and also a lot of misinterpretation of wise advice from leading quality gurus. If you've read this far, you know that the original name for quality control was statistical process control, and that it was designed to replace inspection in certain limited cases where sampling and statistics can be applied. But the word "statistical" got dropped from the name, and we ended up with a confusion. Sometimes, quality control has a very narrow meaning of statistical process control, and sometimes it has a very broad meaning including both inspection and statistical process control, which makes quality control almost synonymous with checking. To make things worse, inspection— an excellent technique everyone should use—got a bad name after QC came into the picture. Let's clarify all of this with a few key points:

- All kinds of checking are great—including both inspection and statistical process control. The more, the sooner, the better.
- The results of inspection and statistical process control should be used not just to evaluate tested products, but also to control and improve processes by making adjustments to our process and applying PDCA for continuous improvement.
- When the requirements for statistical sampling are met, then statistical process control, also called statistical quality control or simply quality control, can be used.
- Other statistical techniques besides statistical quality control can be used for quality assurance, quality auditing, quality improvement programs, and other quality management activities.

To apply statistics to quality management, first, your company has to be ready. Second, your problem has to be one that can be analyzed effectively using statistical methods.

IS YOUR BUSINESS READY FOR A STATISTICAL SOLUTION?

Statistics applies to repeated observation and measurement of similar items. If we don't have procedures in place and are not already doing good quality work, we are not ready for statistics. Here are two examples:

- If our workers currently make ham-and-cheese sandwiches any way they want to, we can't be sure that every sandwich will be good. It would be valuable to standardize ingredients and quantities, then standardize production methods. These quality management activities would give us a consistently high-quality sandwich at acceptable cost produced as quickly as possible. All of that should be done before we try to use statistics. There is no need to draw fancy graphs and curves showing the minimum, average, and maximum number of minutes it takes to make a ham-and-cheese sandwich until after we've standardized the sandwich-making method. Once we have standardized the method, we might apply statistics, as we will see later in this chapter.
- Careful design and administration of surveys or questionnaires is essential. Statistical reports that present the results of a poorly designed questionnaire are garbage—as in garbage in, garbage out. It is actually harder to design good survey questions than it is to apply statistics to the results of a survey. Statistics must always be used in the context of good technical work—in this case, survey design is the technical work that precedes the application of statistics.

QUICK QUALITY TIPS

1911 Before 1924

Make good use of everything Taylor taught in 1911 in *Scientific Management* before jumping ahead to the 1920s technology of Shewhart's statistical process control. To do the best quality management, first standardize procedures—that is, define a good way to do your work, write it down, and ensure everyone follows the same methods—before bringing out the statistical tools.

IS YOUR PROBLEM STATISTICAL?

Statistics applies to large quantities of identical things. We don't even think about using statistics unless we have hundreds of identical items. Many statistical tools are really only meaningful when we have thousands or more identical items. The high end of statistics—Six Sigma—applies to manufacturing situations with millions of identical events. As a result, we don't apply statistics when we are dealing with unique items.

QUICK QUALITY TIPS

If it's Unique, Skip the Statistics

If we make a million identical bricks, and want to make sure they are all strong enough to be used to build our building, we can use statistics to test 10% of them and be sure that all of the bricks are strong enough for the building. But if we are writing a computer program made up of a million lines of computer code—each line unique and different—we can't test 10% of the lines and be confident that the rest are bug free. When each item is unique, we have to test each item—using inspection, not statistical quality control.

So, when we have a standardized process to produce hundreds or more identical items, we can apply statistical process control. Should we? If we've taken care of all the other aspects of quality management, then, yes, we're ready for statistics. But there is one other consideration. Recently, newer technologies have begun to replace statistics with an older method—inspection. Why?

Statistics is important to quality control when the cost of testing every item is high. However, with recent advances in robotics and computing, the cost of measuring every item and checking it against standards has dropped a great deal. When we check every item, we're back to using inspection methods, instead of statistical process control. The field most affected by this is auditing. Auditing used to require a great deal of sampling. Now, the records auditors need to review to test controls are almost always on a computer. Computer data can be inspected by computer programs. The process is so fast that it is less expensive to check every item and skip the sampling and statistics than it is to do random sampling and then calculate statistical results. The same may apply any time that data can be collected automatically by computer. Likewise, it was in the manufacture of electrical (and later electronic) components that Bell Labs developed statistical process control; but with the advent of high-speed robotic manipulation and sensors, it can be affordable to inspect every electronic component instead of checking a sample.

This is not to say that we shouldn't use statistics. Rather, we want to understand the place statistics has, and know when to bring a statistician onto our team to take the next step. Of course, in some environments—especially manufacturing and market research—statistical methods are routinely in use.

Key Statistical Concepts

In this section, you will get a very brief overview of the essentials of statistics. If you are not familiar with these concepts, I highly recommend a book I co-authored, *Business Statistics Demystified,* which will give you a solid grounding that will allow you to understand how to apply statistics for quality management. In this summary, we will cover these issues:

- Measurement for statistics
- Samples and populations
- Descriptive and inferential statistics
- Normal curves and standard deviation (sigma)
- Asking the right question and setting up the experiments

MEASUREMENT FOR STATISTICS

We have already discussed measurement, but we need to go a good deal deeper to define measures that can be used for statistical analysis. In statistical terms, each measurement is taken on one attribute of one subject (for example, our process, our product, or a component), on one occasion (at a particular time and in a particular situation), resulting in a value. We may want to measure many attributes at the same time, as they may affect one another. For example, the amount of humidity in a drying room can affect the dryness, and therefore the weight, of our product. So we would want to record the amount of time the product was in the room, the temperature, the humidity, and the weight of the sliced ham drying on the rack all at once. Every measurement also is associated with a variable that names what we are measuring (weight of ham, weight of cheese, length of hoagie roll) and a unit of measure. The end results of measurement are data—measured values that we record. Data is the input for statistical descriptions and statistical procedures.

What if our measurements aren't good? This brings up the issue of measurement error, which is very different from error in quality management. Error in quality management is about mistakes in production process. Measurement error is about mistakes in the measurement process, which would prevent us from finding the mistakes in the production process. Statisticians talk about two kinds of error: reliability and validity. Reliability is like precision—it is about results being closely clustered together, rather than widely scattered. Validity is about approaching the target, as opposed to being consistently in the wrong direction—validity indicates the absence of bias.

SAMPLES AND POPULATIONS

The idea of samples and populations is key to statistics. A population is the total group in which we are interested. It might be every ham-and-cheese sandwich we ever made or ever will make, or every one we made in a particular factory in a year, or the entire batch of 10,000 ham-and-cheese sandwiches in the fridge that is due to go to the customer tomorrow. A sample is part of a population. Our goal is to measure a sample and be confident of something about the whole population. This can only happen if we are confident that the sample represents the population reasonably well. There are many types of samples, named according to how we select the sample from the entire population. Of all the types of samples on the following list, only the first two give us a reasonable degree of confidence that our sample will represent the population.

- *Comprehensive sample.* In a comprehensive sample, we seek to include the entire population in the sample. The difference between a population and a comprehensive sample is due to mistakes such as missed items or lost data.
- *Random sample.* In a random sample, each item in the population has an equal chance of being included in the sample. Getting a random sample is harder than you think!
- *Convenience sample.* We want to avoid this kind of sampling, where we simply get the sample in the easiest, cheapest way. If we do this, our sample is unlikely to represent the population. For example, if we want to check the quality of sandwiches in our stores, it would be most convenient just to get sandwiches from the nearest store. But they wouldn't be likely to represent sandwiches in all stores accurately, because they were all stored in the same refrigerator at the same time. Very different things might have happened to sandwiches in other stores.
- *Systematic sample.* Here, we get a sample in a non-random way. For example, we might pick the ham-and-cheese sandwich in the top right corner of each box. But what if our refrigeration unit is colder on top and warmer on the bottom? Then we'll never see the moldy sandwiches.
- *Judgmental sample.* This is a term from auditing, actually, rather than from statistics. It means using our own common sense—our expert judgment—to decide how to take our sample. For example, we might choose to look at the sandwiches from only the stores that failed health inspections in the last year.
- *Stratified sample.* Building a stratified sample is complicated, but essentially, a stratified sample is a combination of a judgmental sample, then random items selected from groups selected on the basis of expert judgment.

- *Quota sample.* Again, a poor choice, the quota sample is similar to the convenience sample, except that we stop collecting when our sample is large enough.
- *Self-selected sample.* In this case, the subject has a say in whether or not to be included in the sample. Of course, ham-and-cheese sandwiches can't stand up to be counted, but customers can. A good example of a self-selected sample is customers who choose to answer our customer survey. Unfortunately, we can't be sure that the self-selection doesn't bias the sample.

Just as we can have error in measurement, we can also have sampling error. If we know we didn't use a random sample, then we know that our sample is probably not representative of the population. If we think we have a random sample, then we have more confidence that there is little sampling error and that the sample is similar to the population. However, whenever we can't take a comprehensive sample, there is always some doubt as to the question of whether our sample truly represents the population.

DESCRIPTIVE AND INFERENTIAL STATISTICS

Once we select a sample and take measurements of attributes of items in the sample, we have our data values, which are the input to our statistical calculation. Descriptive statistics are a summary of the data. Commonly used descriptive statistics include the minimum and maximum values, and any of several types of averages (mean, median, and mode, among others). We can also calculate the variance and the standard deviation, which express how our sample is clustered near or spread out far away from the mean (the average or central tendency of the sample). Descriptive statistics describe our sample, but they can also be extended through a process called estimation to describe the entire population. For example, if we try out an advertisement on 1000 people randomly selected from a list of 100,000, and 20 of them buy a ham-and-cheese sandwich, then we have a 2% purchase rate. If we advertise to all 100,000 people, we should expect to sell about 2% of 100,000, or 2000 sandwiches. A good statistician could tell us more; he could make that "about" more precise. Depending on the sample size, he might say, "There is a 68% chance that you will sell about 1900 to 2100 sandwiches."

Inferential statistics go beyond description. Inferential statistics provide a measure of the relationship between two or more variables along with a second measure that indicates how confident we are that the first measure is correct.

Using inferential statistics, we can do things like forecast likely future events and determine if a particular intervention is likely to have a desired effect.

Although estimation and forecasting have more general meaning in business, in statistics, *estimation* always means making statements about the population based on statistics from the sample, and *forecasting* always means predicting expected future measures or results based on past measures or results. In forecasting, it is essential to remember that we can never truly know the future. Rather, we are saying, "If the future is like the past, then this is what is likely to happen." Also, our past numbers do not cause our future numbers. Rather, our past numbers are a result of past causes. Those numbers show the state and trend of the figures. A forecast says, "if the underlying causes remain similar to the past, and the numbers continue to change in the same direction, then the future is likely to be like this."

NORMAL CURVES AND STANDARD DEVIATION (SIGMA)

When we gather the data values from our sample, we can plot them on a curve. Often, the results look like Figure 10-1, the normal curve, also called the bell curve. The normal curve is the representation of a population that is randomly distributed around a central point called the mean. Our results might vary from the normal curve in a number of ways. If the curve leans one way or the other, we say it is skewed. If the curve has two peaks, we call it bimodal, and we know we need to examine our data and data collection methods. Skewed curves are shown on the left of Figure 10-2, and a bimodal curve is on the right. Most statistics won't give useful results if our sample population has a bimodal distribution because most statistics operate on the assumption that the curve of sample

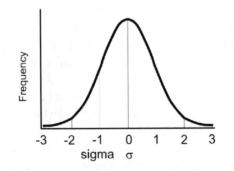

Fig. 10-1. A normal curve.

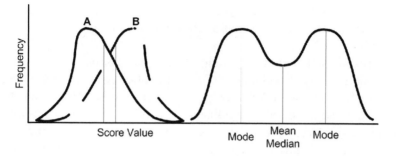

Fig. 10-2.   Skewed and bimodal curves.

and population are normal or near normal—and the math doesn't work right when that's not true.

If we've done our homework—particularly used a good sampling method, chosen an appropriate sample size, and measured well—then the shape of the sample curve will be pretty much the same as the shape of the population curve. Thus, if the sample curve is normal, we can figure the population curve is normal, too. Why does that matter? A statistically normal curve is the result of natural and inevitable variation with no particular cause. A non-normal curve is an indication of the presence of some special cause or causes that are changing the shape of the curve. For example, if we picked a random sample of high school senior boys and measured their heights, we would probably get a normal curve because there is a certain average height, and about equal numbers are taller and shorter than average, but more are close to average than way off average in either direction. But if we picked senior boys playing a pickup game of basketball, and half the kids were on the high school basketball team, we'd be likely to get a skewed curve. If we picked the kids who happened to be in the gym on a particular day, and that turned out to be the basketball team and the wrestling team, we'd be likely to see a bimodal distribution. Now turn all of this around. If we measure a sample or a population, and it's distribution curve doesn't match the normal curve, then there's a reason, and it makes sense to go find out what that reason is. We'll see how to do this later in this chapter.

If the curve is a normal curve, then we can identify the degree of variation from the mean average using an idea called sigma. As shown in Figure 10-1, sigma is a quantity or distance measured outward in both directions from the mean of the curve. One sigma is closest to the central line—the mean; two sigma includes one sigma and more, and so forth. In our sample, more of the values are more clustered near the mean, and that is why the curve is higher at the center. A very important issue in statistics and quality management is the percentage of the total area under the curve within each range of variance, measured in sigma. These values are shown in Table 10-1.

As we become better at quality management, more of our products are within tolerances. Before TQM, 3 sigma (99.75% defect free) was very hard to achieve. However, U.S. business broke through that barrier in the 1980s, and started to move up. Because the curve is nonlinear, percents are no longer meaningful—everything is way over 99% error-free. So we jump from measuring in percent (events per hundred) all the way up to measuring in events per million at 4 sigma. Four sigma quality is 999,936.66 good products per million, or only 63.34 failures. Statistically, 4.5 sigma quality has fewer than 7 failures per million. (In Chapter 13: *Six Sigma,* you'll learn why Six Sigma quality managers call a success rate that is statistically 4.5 sigma "Six Sigma.") By the time we reach statistical six sigma, we are actually looking at fewer than two errors per billion measurable events. That's a lot of quality.

If we know the size of the total population, then we can use a ratio to determine likely quantities of any range of values in the whole population. For example, if our sample of 10,000 1-inch screws shows that 9975 screws are between 0.90 inches and 1.10 inches in length, then we know that 3 sigma (99.75%) of

Table 10-1.   Sigma values and percentages of the sample.

Sigma Value	Range (in sigma)	Unit of Measure	Part Under the Curve and Between the Lines	Part Outside the Range (both sides)
In Quality Management			**Defect-free per Unit of Measure**	**Defects per Unit of Measure**
1	−1 to +1	Percent	68.27%	31.73%
2	−2 to +2	Percent	95.45%	4.55%
3	−3 to +3	Percent	99.75%	0.25%
4	−4 to +4	Per million	999,936.66	63.34
4.5	−4.5 to +4.5	Per million	999,993.20	6.80
5	−5 to +5	Per million	999,999.43	0.57
6	−6 to +6	Per billion	999,999,998.03	1.97

our sample fall within that range. If we produce a million screws, we can expect about 997,500 screws to fall within the same range.

ASKING THE RIGHT QUESTION AND SETTING UP THE EXPERIMENTS

Earlier, we laid out the process of taking samples and measurements. Our actual work will go in this order: select a sample; take measurements; do our statistics; use the statistics to answer business questions. However, it is essential that our planning go in the opposite order. First, we have to define the business question or questions we want answered. Then we explain that question to a statistician, who determines what statistical procedures we will need, and then how to select the sample and gather measurements necessary for those procedures. If we start gathering data before we have done our planning, there is a very good chance that we won't collect the right data, and that we'll have to start all over again.

STATISTICAL SIGNIFICANCE AND BUSINESS SIGNIFICANCE

When statistics is applied to science, there are very specific rules for statistical significance—for determining if statistical results provide enough information to support a new theory in place of an old one. However, this is not true for business. In business, we have to work closely with definition of business value in relation to statistical results so that the statistics are properly applied in making a business decision. Also—in a way similar to what was said about findings in the audit chapter—statistical significance and business significance may be at odds with one another. For example, one time, an auditor told me how frustrated she was because she was looking for any significant correlation between methods used to do a particular type of work at different locations and effectiveness of the work. She couldn't find any correlation that was statistically significant. I pointed out that the lack of a significant difference was a very significant business result, because if all methods are about equally effective, then the organization can save money by standardizing on the least expensive method. Statistical analysis showed us no significant difference in effectiveness—which means that the cheapest method is just about as good as the most expensive. A smart businessman will use a less expensive solution if it's just as good as a more expensive one. Statistical insignificance can be significant for business!

Summary of Statistical Techniques for Quality Management

In this section, you will learn about the most important and commonly used tools of statistical quality management. For an in-depth discussion of these tools, their variations and many more statistical tools used in quality management, read *Six Sigma Demystified* by Paul Keller. Here you will learn to find root causes with Ishikawa diagrams and prioritize errors for elimination with Pareto optimization. After that, we will introduce the statistical quality control chart and the concept of seeding errors to determine the success of inspection.

FINDING ROOT CAUSES AND ELIMINATING ERRORS

These two tools are commonly used in Total Quality Management, and have been adopted by other methods.

Ishikawa—Cause-and-Effect—Diagrams

The Ishikawa diagram, also called the Cause-and-Effect diagram, or the Fishbone diagram—because of its shape—is used to identify multiple causes of a single result. It is an excellent tool for PDCA and the design of experiments. It works best as a team tool, built from brainstorming sessions where everyone suggests possible causes of a known error, problem, or defect.

Figure 10-3 is an Ishikawa diagram for an upscale ham-and-cheese problem. Let's say that we run a restaurant—a French bistro—famous for its croque monsieur, the elite French version of the ham-and-cheese sandwich. (For a recipe, go to www.qualitytechnology.com/QMD.) Our restaurant has recently learned that we have some customer complaints that they ordered the croque monsieur, but did not get it. This is of concern, because, next month, our restaurant is being written up in a major travel magazine as the destination for croque monsieurs. We want to make all of our new customers happy. So, we need to find and eliminate the causes of the missing croque monsieur.

We begin by setting up the Ishikawa diagram with a long central arrow and the effect, "no croque monsieur" on the left. Each of the major angled arrows—called first-level arrows—represents a category of causes. We can use any categories we want. In this case, we have chosen to use a standard, the Five Ms used in the

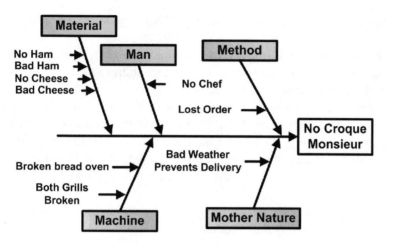

Fig. 10-3. The Ishikawa, or Cause-and-Effect diagram for unavailability of ham sandwiches.

manufacturing environment: material, man, method, machine, and mother nature. If we wanted to focus on customer service or transactional issues, we could use the Five Ps: people, places, policies, processes, and procedures. Or we could use our own organizational chart, flow diagrams, or anything else. And we can add new categories whenever we think of them. The purpose of the categories is to help us find all of the causes, which are the smaller arrows pointing towards the angled first-level arrows.

To find causes, we can look at records—either internal documents such as food orders, or customer information, such as complaint cards—and we can also sit down and try to remember what has gone wrong. The more people we get involved, the better. In this case, we've identified four materials problems—ham or cheese can run out, or it can be less than perfect (bad). We have one man-power problem—only the two senior chefs have perfected their croque monsieur. If both of them are out, there is no one else who can make it. Occasionally, there is a failure of recording or communications methods, and an order gets lost. One time, the bread oven broke down, and our custom baguettes couldn't be made. (Nothing else will do for a croque monsieur.) Once, the grill broke. And a bad storm can keep supplies from coming in.

We gather as many reasons as we can, post them on the Ishikawa diagram, and then look for more. We can also document how many times each problem occurred in a particular time period. When we feel we've found every cause, we're ready to plan corrective action.

How much corrective action is needed? That depends on two things:

- Are we aiming to eliminate the problem entirely? If so, we will have to address all the causes. Otherwise, we can address the biggest reasons and work our way down from there.
- Are there any situations where two different causes must both happen to create the problem? For example, if we have two bread ovens, then, as long as one is working, the croque monsieur is safe. So we don't have to get a zero-defect bread oven to get a 100% available croque monsieur; we only have to make sure that at least one bread oven is always working.

Making regular use of Ishikawa diagrams and PDCA will teach you these three things about quality management:

- *Almost any problem can be solved if you break it into pieces.* Problems can feel overwhelming until we define and diagnose them.
- *We should try to solve problems, not work around them.* We aren't considering a menu alternative or a gift certificate to manage customer dissatisfaction. We're not looking just to know we have a problem so we can say "sorry, monsieur, no croque monsieur." We're seeking to keep the problem from ever happening again.
- *There are only three things that make a problem unsolvable.* Either there really is no answer, yet. Or we don't have the time, people, and money to solve the problem. Or—most often—the corporate culture won't allow a solution. This is why good quality management initiatives seek to discover and create best practices, to allocate resources for quality solutions, and, most importantly, to transform corporate culture.

Pareto Optimization

Our next tool, the Pareto Diagram—also called a Pareto chart—is fun because, although it depends on a statistical law, we don't have to do any statistics to use it. We just have to organize our data in a certain way. We often use the Pareto chart after the Ishikawa diagram because the results of cause-and-effect diagramming can be used to build the Pareto chart. If we look at Figure 10-3, we see that there are nine causes of not getting a croque monsieur to the customer. Instead of seeing those as nine causes of one effect, let's think of them as nine cause-effect pairs. A *cause-effect pair* is a unique pair of events that describe a cause that results in a defect of product or service. Of course, each event pair can happen many times.

To prepare the Pareto chart, we gather data about defect events, organize it into cause-effect pairs, and count the number of times each cause-effect pair occurs. We might find this data in help desk logs, where each incident (not each call, but each set of one or more calls related to a single user problem) is logged. When diagnosed, it is a cause-effect pair. Or we might gather the results of statistical quality control or inspection. Any data about a large quantity of defect events on a single set of processes can be analyzed by creating a Pareto diagram.

In this case, let's say that our success with finding causes for missing croque monsieurs leads us to try to solve every case of customer dissatisfaction at our bistro. We perform surveys, offer free dinners to anyone who complains, and do all we can to gather lots of data. We sort it by type of defect: item on menu not available; wrong item delivered; food too hot; food too cold; food too late; food not to taste; food not what the customer expected. For each of these effects, we can use an Ishikawa diagram or another tool to find all the causes. From all of this, we build our list of cause-effect pairs, and then put it in order with the one that happens most listed first, and on down from there. The total number we get from adding up the number of times each incident happens equals the total number of customer complaints. We then order the list by type of defect, with the most common problem first, then the second most common, and on down.

When we do this, we will discover an amazing fact—the illustration of Pareto's law. No matter what situation we look at, the distribution of the frequency of cause-effect pairs is always just about the same. Eighty percent of all defect incidents are caused by only twenty percent of the sources of problems. This is illustrated in Figure 10-4.

This is very good news. When Pareto's law applies, it means that, if we prevent the cause of only 20% of the cause-effect pairs, we reduce the number of defect incidents by 80%. In one situation, I helped analyze help desk call logs. There were about 525 calls per week—way too many. We did a Pareto analysis,

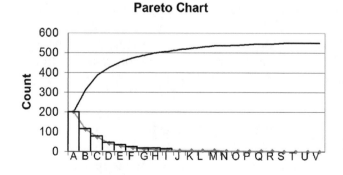

Fig. 10-4. A Pareto diagram.

then had technical teams come up with permanent solutions for the first five of the 22 different cause-effect pairs we found. We were down to 125 calls per week. We did it again, and we were down to 25 calls per week. At that point, the help desk could easily handle all the calls. If we'd done another iteration of analysis and permanent preventative solutions, our help desk staff would have been as lonely as the Maytag repairman! In fact, iteration of corrective activity using Pareto diagrams is a way to move quickly to quality levels of three, four, or five sigma. (Six sigma takes some heavier-duty statistics, as you'll see in Part 3.)

Deeper Root Cause Analysis

There is one other step we can take with the information from Ishikawa and Pareto diagrams. Before we try to solve each cause, we can ask, "Is there a deeper pattern here? Is there a single cause of many causes?" Let's say, for example, that I examine the causes of drops in my personal productivity. I take notes each hour for a week catching anything that reduced how much work I got done. By the end of a week, I have quite a list. Perhaps I find 20 or 30 cause-effect pairs. Then I take a closer look. I might ask about each item, "Why is this happening?" and then ask "Why?" of that answer again, five times. If we go deeply enough, we can get to a root cause. Sometimes, many of the original causes on our list all have one root cause. For example, I see that my list includes phone calls from three different people, office visits from five others, and looking at my email when that tinny voice says, "You've got mail." I start asking, "Why is this happening?" I say, "Because I'm being interrupted." I ask, "Why am I being interrupted?" A light bulb goes off in my head. All of these different things—phone calls, visits, and emails—have one cause: I'm letting myself be interruptible! So I make one change—I stop being interruptible—and all those cause-effect pairs go away. That one change may require several steps: I turn off the ringer on my phone; I close my office door and put up a *do not interrupt* sign; I let people who contact me frequently know I'm not available; and I change my email settings to download only on my command. All these actions have one purpose, and I resolve many causes of the problem in one stroke. I'm like the little tailor who killed seven in one blow!

THE QUALITY CONTROL CHART

Now we will turn our attention to a tool that has been used to change manufacturing processes around the world—the statistical process control chart, also called the quality control chart, illustrated in Figure 10-5.

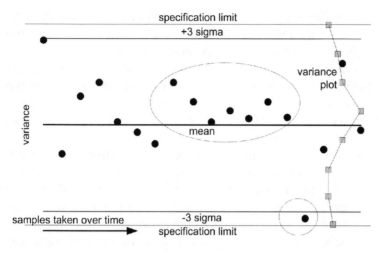

Fig. 10-5.   A basic quality control chart.

Elements of a Quality Control Chart

A quality control chart is used to display data sampled over time. Think of a process—perhaps a machine or an assembly line or a computer reading checks and recording transactions—steadily producing output. We take a sample of that output, and measure a variable. The results shows up as a dot on the control chart, with more dots appearing over time. The vertical location of the dot above or below the centerline—the mean if the process produces a statistically normal distribution—shows the variance of that particular sample from the target mean. The diagram is bordered at the bottom and top by two pairs of lines. The outer pair is called the specification limit and represents the outer limits of the tolerances defined in the specification for this product. That is, the range between the outer lines is the ideal specification (the center line) plus or minus the tolerance. Inside those two lines, we see another pair of lines which, in this diagram, are marked "3 sigma." These are the control limits—the upper control limit (UCL) and the lower control limit (LCL). We set the control limit so that, if the sample remains within it and certain other criteria are met, we are highly confident that the entire population—everything we make—is within the specifications limits. In this way, we can test only a sample of the output and be confident that the entire output is within the specification limits and is acceptable to the customer. That is one of the purposes of the control chart.

If we use the chart only for that purpose—to show if the production run is good or not—then it isn't really being used to control anything, it's just being used for testing. But we can also analyze the chart and investigate certain points that indicate a problem with our process. If we correct the causes of those problems, then the process will come back under control.

Essential to this idea is the underlying principle of the control chart—that variance from the mean comes from one of two types of causes. One type, called a common cause, is not an indication of a problem. It is just normal variation, and it can't be eliminated. The other type—called a special cause—can be considered an error in the process.

A key point made over and over by Deming and Shewhart is that there will always be variance of both quantity and quality of output due to common causes. As a result, it is a mistake to penalize people for doing less than average work or praising them for above average work. It is also a mistake to go chasing after phantom problems if you have not determined that a process is truly out of control. Adjusting a process that is operating within normal variance is called tampering, and it interferes with the process and the workers.

On the other hand, the presence of a special cause indicates that our process is out of control. Our job will be to find the cause and correct it, improving the process so that the process is once again under control, the sample is within limits, and the product meets specification. There are five patterns on a control chart that indicate a process is out of control. The first two appear on Figure 10-5 and are circled there.

- One point outside the control limits.
- Seven points on one side of the mean.
- Seven sequential ascending or descending points.
- If two out of three points are on the same side of the mean and outside two standard deviations.
- If four out of five points are on one side of the mean and more than one standard deviation from the mean.

Why are these particular patterns indications of an error? Because we can show statistically that any of these events is highly unlikely in a normal distribution. So, if one of these events is happening, the distribution is not normal and out of control. And there must be a special cause for that.

The linked set of squares at the right side of the chart is called a variance plot, and it is a visual summary of the dots on the chart. If the process is in control, it will approximate a normal curve. As you can see, the two cases of special causes show up as distortions of the variance plot away from a normal distribution.

The information from a control chart feeds into PDCA, including perhaps an Ishikawa diagram to find out the special causes, or a Pareto chart to identify the priorities—the most common or most costly sources of out-of-control conditions. Then we take remedial action and test it by seeing what shows up on the control chart.

There are many variations on the control chart to handle special circumstances, such as varying sample size and a need to group together samples taken over time. There are also special uses of control charts. For example, we can design experiments to vary one variable at a time, or make small adjustments on several variables, and then use a control chart to see what happens to the process. These can be used to hunt down difficult special causes, to optimize a process, or to vary the mean to suit a change of requirements.

What happens when we bring all of our processes steadily under control? Where do we go from there? We don't need to sit on our laurels. If we do, we are just sitting around putting out fires, keeping things under control, and not improving. Deming compared this to responding to a fire alarm at a hotel by putting out the fire, saying "Extinguishing the fire does not improve that hotel."

Once our process is under control, we can improve it in one of two ways. One approach is to tighten our controls. Instead of just having a 3 sigma range as our control limit, shooting for a quality level around 99.75%, we can try to tighten the process so that 4 sigma, or 5, or 6 becomes our control limit, and our defect rates drop to just a few per million or per billion. When we reduce the errors to extremely low levels, we don't talk about percentages any more. We begin to talk about defects per million opportunities (DPMO). An opportunity is a condition or a moment in the process where a defect-producing error or a defect in the product—interpretations vary—could occur. This is the realm of Six Sigma quality engineering, which we will discuss in Chapter 13.

The other alternative is to re-invent our process with new technology or outside-the-box thinking. We can invest in research not just to solve problems, but rather to make radical improvements. If we don't, someone else probably will, and the technological base of our industry will shift under us before we know it. This doesn't mean that every company has to invest in its own version of the Palo Alto Research Center or Bell Labs. We can also research innovation in our industry and related industries, and think of new things worth doing in our company.

SEEDING ERRORS

The final tool I want to introduce actually predates statistics, but is also used for the very highest level of zero-defect quality engineering. The method is called seeding errors, and it consists of intentionally putting errors into the process to see if they are caught by our inspection or quality control efforts. If they aren't caught, that means that we aren't catching all the errors. Taylor mentions a very sophisticated use of seeding in his 1911 treatise *Scientific Management.* He used seeding—planting ball bearings known to be defective into certain batches—to determine how effective human inspection of the product was. He used this to perform experiments, such as testing how often people should take breaks to maintain maximum effectiveness. Seeding errors is still used, because it can help us answer a difficult question: How do we know when there are no errors left to be found? I would argue that there is no way to be absolutely sure, no way to prove that something is error free. But we can define tests that make it extremely unlikely that there are any errors left, and one of them is seeding.

If we seed a significant quantity of errors of each known type into our batch of product, then have the team re-inspect the product, we can see how quickly they find all of those errors. Measuring the effort needed to find a known quantity of errors allows us to extrapolate how much effort it would take to find all the errors. If we do that much, plus a margin of safety, moving well beyond the point in time when the last error was found, we can make a case that we are likely to have reached zero defects.

Some people see problems with this approach. For example, what if the error that wasn't found is a completely different type of error that never occurred to anyone? If we didn't know that type of error existed, we couldn't seed with it. Also, in physical products, we can seed with actual defective products that have been caught by inspectors by pulling them out of the discard pile. But in a realm like software inspection, we have to make up errors that we believe are the same type as errors we're looking for. But what if the error looks the same as a real error to us, but not to our inspectors? That could invalidate our test. Lastly, there is a practical concern. We need to make sure that we remove all the errors we seeded before we send the product to the customer.

Even given these concerns, we can say two good things about seeding errors. One is that—in trying to answer the difficult question of whether any errors remain—seeding errors and seeing what our inspectors do is better than doing nothing. The second is that it is part of a process that works. It has been used in the context of software inspection, which has, indeed, produced software that was found to be free of defects in production.

The Statistical Quality Team

Before we close, I want to reconnect the technical side of quality with the human side. Statistical quality management is done by people, and its success depends upon human motivation and effective communication. The statistical quality team is perhaps more highly trained and more specialized than other quality teams. If anything, that demands even better people skills. Statisticians and engineers cannot allow their ability to do sophisticated work and speak in technical jargon to reduce their ability to speak to—and, even more importantly, listen to and understand—customers and non-technical stakeholders. Training and leadership in statistical and technical expertise must be balanced by training and leadership in communication and commitment to customer quality.

Conclusion: Statistics Enhance Quality

There are two situations where statistical techniques and related analytical tools are very helpful in the work of quality management. One is when things are very, very bad, and the other is when things are very, very good. When things are very bad, we can feel overwhelmed by the number of problems and their complexity. In that situation, tools like Ishikawa diagrams to find causes and Pareto charts to prioritize our problems help bring error under control. We can tackle the most common or most costly errors first. These cost- and time-saving methods can be fed back into the continuous improvement effort. Iterating our improvement effort, we can bring a very bad situation under control.

We also need statistics if we're doing a very good job and we want to do better. As we improve the quality of our processes and products, there are fewer and fewer errors. At a certain point, errors simply become very hard to find. Also, obvious causes are all taken care of, and we need subtle analysis and carefully designed experiments to find subtle causes. At this point, we're going to need statistical techniques to hunt down the error and their causes so that we can keep improving.

QUICK QUALITY TIPS

Statistics at the Best of Times and the Worst of Times
Remember this:

- When our problems are simple and obvious, just fix them.
- When our problems are overwhelming, use statistics to understand them and know which ones to hit first.
- When our problem are few, use statistics to dig them out and get rid of them.

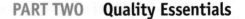

Q-Ball Quiz for Chapter 10

1. Which of these is *not* essential to do before using statistics?
    (a) Standardize operations.
    (b) Create quality teams with good communications.
    (c) Set up an internal audit department.
    (d) Prepare clear specifications.

2. An Ishikawa diagram
    (a) helps us find many causes for one effect.
    (b) helps us find the many effects of one cause.
    (c) helps us detect defects.
    (d) helps us detect defectors.

3. Which of these jobs would probably *not* be done with a Pareto chart?
    (a) Prioritizing problems to solve first.
    (b) Trying to get a handle on an overwhelming number of customer complaints.
    (c) Deciding which special causes from a control chart to tackle first.
    (d) Trying to determine if we have found every defect in a product.

4. Which of these is *not* an indicator of a special cause on a control chart?
    (a) A point outside either of the control limits.
    (b) Four sequential ascending or descending points.
    (c) Seven sequential points on one side of the mean.
    (d) Two out of three points are on the same side of the mean and outside two standard deviations.

5. Variations on control charts handle all of these cases *except*
    (a) varying batch sizes.
    (b) higher quality, with more levels of sigma within the requirements specification.
    (c) special adjustments to ensure we have found all the defects.
    (d) adjustments for samples taken over time.

Mid-Term Exam

1. A best practices is
   (a) a high-level guideline to use in project planning.
   (b) a universal principle that cannot be broken.
   (c) the best way to do a repetitive process.
   (d) out-dated and no longer used in quality management.

2. Synergy is
   (a) the quality of people being able to do more together than we could
       do separately.
   (b) a meaningless buzzword.
   (c) implemented through a carefully structured plan.
   (d) of little real value.

3. Effectiveness is
   (a) our ability to achieve desired results.
   (b) our ability to minimize waste.
   (c) the same as productivity.
   (d) less important than efficiency.

4. A destruct test
   (a) destroys all the evidence of an error.
   (b) should be used on every product we produce.
   (c) destroys the product, but gives us information about defects in the batch of products from which it came.
   (d) is a really bad idea.

5. In quality management, objective means
   (a) the same, no matter who does the checking.
   (b) of, or relating to the object, that is, the product.
   (c) objectionable.
   (d) giving no reason for anyone to object.

6. A process requirement is
   (a) a requirement on a defined measure of a process that is happening.
   (b) only used to measure processes that have no effect on output.
   (c) only used in service industries.
   (d) a made-up term that doesn't mean anything.

7. Change Control requires all of the following *except*
   (a) defining change requests.
   (b) updating specifications.
   (c) communicating change to team members.
   (d) deciding what to change.

8. Forecasting
   (a) is impossible, because we can't predict the future.
   (b) means predicting expected future measures or results based on past measures or results.
   (c) is a form of estimation.
   (d) is good, but is not as good as Six Sigma fivecasting.

9. Subjective risk is
   (a) a risk gathered from customer polls or other subjective sources.
   (b) a risk that has not been objectively measured.
   (c) a risk that can't be objectively measured.
   (d) a made-up term.

10. The voice of the customer is best described as
    (a) a process of including customer communications in specification development.
    (b) a catch phrase to remind us that we need to be certain we are addressing customer requirements.
    (c) a technique for ensuring customer satisfaction.
    (d) the most important element of Six Sigma.

11. Quality Assurance focuses on
    (a) improvements to design and to work process that eliminate error.
    (b) reducing the cost of finding defective items.
    (c) ensuring work is done according to standard.
    (d) increasing the number of defects found.

12. It is possible to be too customer-focused. When we are,
    (a) we deliver quality to the customer, but don't pay enough attention to the bottom line.
    (b) we take care of the customers, but ignore vendors.
    (c) we take care of the customers, but don't improve our own processes.
    (d) we take care of the customers, but ignore our own workers.

13. A requirements tracing matrix does all of these *except*
    (a) link a requirement to its source.
    (b) link a requirement to components needed to meet the requirement and their relevant attributes.
    (c) link a requirement to the reviews, inspections, and tests that ensure we meet the requirement.
    (d) link a requirement to similar requirements in other products or services.

14. In quality management, a cause-effect pair
    (a) is a unique pair of events that describe a process and its results.
    (b) is a unique pair of events that describe a cause that results in a defect of product or service.
    (c) is a unique pair of events that describe a risk and its consequences.
    (d) is a unique pair of events that describe a vendor and a bad input.

15. Inputs
    (a) are data, not physical items.
    (b) are used up in a process.
    (c) are the same as outputs.
    (d) are ingredients that go into a process to be transformed into outputs.

16. Benchmarking is
    (a) a defined measure of productivity based on comparison to similar processes.
    (b) a way of putting marks on product on the workbench to isolate defects.
    (c) marking time without getting work done.
    (d) a basic requirement of ISO 9000.

17. Statistical Process Control is now generally referred to as
    (a) Statistical Quality Control.
    (b) Quality Assurance.
    (c) Quality Control.
    (d) Statistical Quality Management.

18. A Standard Operating Procedures (SOP)
    (a) is a written technique describing how we do a process now.
    (b) is a written technique describing how a process is done to a general standard.
    (c) is always a best practice.
    (d) must be up to standard, but does not always need to be written down.

19. Which of these techniques does not have as one of its primary purposes the reduction of subjective decisions?
    (a) Fact-based decision making
    (b) Cost of quality analysis
    (c) Pareto optimization
    (d) Brainstorming

20. Efficiency is
    (a) the same as effectiveness.
    (b) our ability to achieve results with less effort or cost.
    (c) the most important factor in quality management.
    (d) the only thing auditors can easily measure.

21. Flow diagramming
    (a) can be used to show how materials and/or information flows through a company.
    (b) replaces flow mapping.
    (c) is only used in computer programming.
    (d) was invented by Six Sigma engineers.

22. An output requirement is
    (a) a requirement of an attribute of the final product exclusively.
    (b) a requirement of an attribute of the output of any process.
    (c) more important than an input requirement.
    (d) more important than a process requirement.

23. Which of these statement is *not* true of Quality Assurance (QA)?
    (a) It is cross-functional.
    (b) It includes redesigning products.
    (c) It includes redesigning processes.
    (d) It includes Quality Control.

24. A sample where each item in the population has an equal chance of being included in the sample is called a
    (a) stratified sample.
    (b) random sample.
    (c) comprehensive sample.
    (d) convenience sample.

25. Tolerance defines the acceptable limits of technical standards as defined by the
    (a) customer requirement, stakeholder requirement, standard or regulation.
    (b) producer.
    (c) technical standards agency.
    (d) government.

26. Fiduciary risks are all of these, *except*
    (a) a part of doing business.
    (b) a risk related to others.
    (c) a risk related to our legal responsibilities.
    (d) entirely internal.

27. A process
    (a) is the activity of changing inputs into outputs.
    (b) is always precisely defined.
    (c) always meets standards.
    (d) is not as important as a product.

28. Requirements elicitation includes all of these *except*
    (a) presenting ideas to customers or customer representatives.
    (b) receiving suggestions for features or improvements from customers.
    (c) identifying all standards that have requirements we must meet.
    (d) recording the results of customer meeting.

29. Accuracy in measurement should be
    (a) as high as we can afford.
    (b) sufficient to ensure that the attribute is within tolerances.
    (c) biased.
    (d) imprecise.

30. In a business environment, which of these is *not* an element of the customer?
    (a) Purchaser
    (b) User
    (c) Consumer
    (d) Corporation or business paying for the item

31. Statistical Process Control includes all of these *except*
    (a) examination of a sample, rather than a whole batch.
    (b) identification of defects.
    (c) innovative thinking to prevent defects.
    (d) a source of information to be used for process improvement.

32. A convenience sample is
    (a) inexpensive, but unreliable.
    (b) inexpensive, and good enough for most statistical work.
    (c) usually biased by self-selection.
    (d) rarely used.

33. Quality Planning (QP)
    (a) includes planning for QC, planning for QA, and more.
    (b) replaces QA.
    (c) replaces QC.
    (d) is outmoded.

34. Which of these statements is *not* true about customer value?
    (a) Customers seek that which they perceive to be the highest value for the price.
    (b) Customers always seek the highest quality, regardless of cost.

(c) Customer perception of value in a product affects customer perception of the value of the producer of that product.

(d) Quality management helps increase customer value in several ways.

35. Which of these statements is *not* true about good design?
    (a) Good design ensures that the product performs its desired functions.
    (b) Good design is only related to aesthetics and has nothing to do with function.
    (c) Good design depends on cultural values.
    (d) Good design takes cost into account.

36. Outputs are all of these *except*
    (a) the end results of a task, such as a component or a finished product.
    (b) inputs to subsequent processes.
    (c) inputs to preceding processes.
    (d) measured by inspection or statistical quality control.

37. Statistical Quality Control does all of these *except*
    (a) helps insure that the product is within acceptable standards.
    (b) helps discover flaws in the production process.
    (c) guarantees that all inferior products are thrown out.
    (d) makes it possible to ensure the quality of production without checking every product.

38. Liability
    (a) is a cost we will, or might, have to pay in the future.
    (b) is only related to fiduciary risk, not internal operations.
    (c) is best managed with insurance.
    (d) is a type of inefficiency.

39. We want our measurement to be all of these *except*
    (a) accurate.
    (b) biased.
    (c) unbiased.
    (d) precise.

40. A part of PDCA is correctly defined on each of these lines *except*
    (a) plan. Define the problem or need, and decide what to do about it.
    (b) do. Put the new process in place.
    (c) check. Take a look at what happens.
    (d) authorize. Authorize a change to the standard to permanently solve the problem.

41. A systematic sample seems good, but it runs into trouble because
    (a) it allows too much self-selection.
    (b) it isn't comprehensive enough.
    (c) it is too random.
    (d) it can consistently miss a systematic error.

42. Customers have different definitions of exactly what gives the most value. As producers, we effectively deal with this using all of these methods *except*
    (a) deciding what the customer will want ourselves.
    (b) making one product, and providing sufficient value, if not the highest value.
    (c) making multiple products in the hope of giving more value to more customers.
    (d) letting our customers make choices about what they want, and then delivering customized products.

43. Tools are best defined as
    (a) things which are used for a task, but not used up.
    (b) things which are used up in a task, but don't become part of the product.
    (c) things that, through process, become part of the product.
    (d) instructions for completing a process.

44. Quality management requires
    (a) application of universal principles, uniquely adapted to particular circumstances.
    (b) discovery of new universal principles.
    (c) precise definition and absolute proof of universal principles before they can be used.
    (d) that we disregard universal principles, because the concept that they exist is faulty.

45. Customer service on the team is all of these *except*
    (a) the idea that, if your process uses my outputs, you are my customer.
    (b) a part of the process of definition of customer requirements.
    (c) a valuable tool for teambuilding.
    (d) a way to link together the SIPOC chain in everyone's mind.

46. Pareto optimization is primarily used to
    (a) identify root causes.
    (b) identify defects.

(c) prioritize defects for the prevention effort.

(d) calculate the cost of quality.

47. Vulnerability to lawsuits, criminal charges, loss of reputation, or other costs is called

(a) inefficiency.

(b) bad karma.

(c) exposure.

(d) poor quality.

48. Root cause analysis finds

(a) the basis that supports quality, the way roots support a tree.

(b) the first process in a complex manufacturing chain.

(c) the one source of all of our problems.

(d) the most basic, original source of a defect or error.

49. A permanent preventative solution is all of these *except*

(a) the solution to a root cause.

(b) a process change that will save money.

(c) an alternative to root cause analysis.

(d) a process change that should become the new standard.

50. All of these statements are true about resources *except*

(a) resources are used up in the process, but don't become part of the product.

(b) effort, or work time, is a resource.

(c) inputs are one type of resource.

(d) reducing use of resources per unit of output reduces the cost of a process.

51. Ishikawa diagrams are also called

(a) histograms.

(b) cause-and-effect diagrams.

(c) flow charts.

(d) Pareto diagrams.

52. The father of Scientific Management is

(a) Fredrick Winslow Taylor.

(b) Walter Shewhart.

(c) W. Edwards Deming.

(d) Philip B. Crosby.

53. The development of Scientific Management included the development of all of these *except*
    (a) statistics.
    (b) accurate data gathering.
    (c) an analytic approach to improving workflow.
    (d) a cooperative relationship between management and workers.

54. Testing is
    (a) always essential before final product delivery, to avoid sending defects to customers.
    (b) the process of actually doing something with a product, service, or component and seeing what happens.
    (c) less expensive than inspection.
    (d) always multiple choice.

55. Corrective action is
    (a) the best approach to quality management.
    (b) what we do after a defect is found.
    (c) is always essential in the action of the PDCA cycle.
    (d) doesn't apply to the results of testing, only to the results of inspection.

56. All of these are true of architecture *except*
    (a) architecture involves seeing from multiple perspectives.
    (b) architecture applies only to the building and construction industry.
    (c) architecture must work within constraints.
    (d) poor architecture will increase the cost of quality.

57. All of these are true of techniques *except*
    (a) techniques are the instructions of the work process, or the way of doing the work.
    (b) when techniques are not standardized, we can observe all current techniques to create a technique from the best steps of each one currently in use.
    (c) it is best for techniques to be standardized.
    (d) techniques can be used to replace processes.

58. We can use all of these as inputs for creation of a best practice, *except*
    (a) existing standards.
    (b) industry best practices.
    (c) PDCA.
    (d) the way we do things now.

59. Input requirement are all of these *except*
    (a) requirements that we might use to check material or components arriving from a vendor.
    (b) requirements specified for the input to a process.
    (c) less important than output requirements.
    (d) one of three major categories of requirements.

60. A quota sample
    (a) is a type of convenience sample.
    (b) is a type of judgmental sample.
    (c) is a type of stratified sample.
    (d) is a made-up term.

61. The inventor of statistical quality control is
    (a) Fredrick Winslow Taylor.
    (b) Walter Shewhart.
    (c) W. Edwards Deming.
    (d) Philip B. Crosby.

62. The work environment
    (a) is an important, but often overlooked, factor in quality management because variations in the work environment can reduce quality.
    (b) is not an important part of quality management.
    (c) is important only where it concerns the safety or productivity of the workers.
    (d) cannot be adequately defined for measurement.

63. Internal auditors have which attitude towards leaving tools behind for management?
    (a) It would be good, but if they do, it compromises their independence.
    (b) Measurement tools left behind are a good way for auditors to add value by making it easier and less expensive for management to improve future work results.
    (c) Management should define the tools.
    (d) An auditor is just like a plumber; he shouldn't leave his tools behind.

64. The measured or assessed level of performance of how things are or were done is called
    (a) a cause.
    (b) a condition.
    (c) an efficiency.
    (d) a result.

65. The consequence of an error is all of these *except*
    (a) what would happen if the error went undetected.
    (b) an important factor in measuring the cost of quality.
    (c) always measurable.
    (d) often borne by the customer.

66. Criteria is best defined as
    (a) an audit or legal standard.
    (b) what should or could be—for example, the relevant standard or best practice.
    (c) an objective basis for a decision.
    (d) what auditors try to find through research.

67. The key figure in Total Quality Management is
    (a) Fredrick Winslow Taylor.
    (b) Walter Shewhart.
    (c) W. Edwards Deming.
    (d) Philip B. Crosby.

68. The inventor of the Cost of Quality approach is
    (a) Fredrick Winslow Taylor.
    (b) Walter Shewhart.
    (c) W. Edwards Deming.
    (d) Philip B. Crosby.

69. Quality Control (QC), in its broader sense
    (a) is almost a synonym for checking.
    (b) is part of Quality Assurance.
    (c) replaces Quality Assurance.
    (d) always uses statistics.

70. Quality management addresses all of these standards *except*
    (a) environmental standards.
    (b) government standards.
    (c) international Standards.
    (d) the Standard & Poor's index.

PART THREE

Quality Movements

In Part 3, we survey each of the quality management movements that has developed since 1950, all of which are still very much active and alive. Having read this far, you now know enough to evaluate each movement, to understand its strengths and weaknesses. If you are launching a new quality improvement program, you will learn enough to choose methods which are best for your company. If you are currently using one of these methods, you will be able to evaluate how well you are using it relative to its innate strengths and weaknesses, understand the pitfalls others have run into, and improve your company's quality management performance—measured in customer delight and bottom-line dollars.

Quality Management has been difficult to understand because, before 1980, it was an obscure field in North America and, since 1980, it has been a contentious one. Although advanced quality management methods have been available since the 1950s, industry in the United States did not think quality was important until the late 1970s. In the first 30 years after World War II, Henry Ford's methods were uncontested because, as far as any executive could see, they worked. And the rapid growth in consumer demand was met by focusing on productivity—getting lots of products to the customers—not quality—making sure that the products were good.

In the meantime, Japan recovered from the destruction of its industrial infra-structure in a different way, by focusing on quality. The approach, later called Total Quality Management (TQM), developed in Japan from the late 1940s through the 1980s. As we've demonstrated, a focus on quality leads to lower cost and higher productivity, as well. By the late 1970s, Japanese manufacturers could produce photocopiers, ship them to the United States, and sell them for a retail price lower than the base manufacturing cost of an identical copier made by the Xerox corporation here in the United States. And that wasn't due to the lower cost of labor or the foreign exchange rate. It was due to quality manage-ment, which eliminated the costs of rework and scrap. Around 1980, North American manufacturers had a rude awakening—they could no longer keep pro-ducing low quality—the cost was too high.

This led to the introduction of Total Quality Management in the United States. The new era was marked by the Ford Motor Corporation hiring W. Edwards Deming to re-engineer the entire company. Most of U.S. industry quickly followed suit, and TQM was all the rage. Some succeeded, and decided to move beyond TQM to a higher goal called Six Sigma. In the arena of military contracting, a parallel movement arose called the *Zero Defect* movement or the *Cost of Quality* movement.

What is rarely seen is that these different methods are really not very differ-ent at all. All of them use the set of methods we introduced in Part 2: defining, planning for, controlling, assuring, and delivering quality. Each emphasizes dif-ferent aspects of these tools, adds some refinements, and couches the problems and solutions in slightly different language. There is a lot more apparent conflict than there is real disagreement. When I think about the conflicts among schools in quality management, I am reminded of the painters Monet and Picasso. Both were incredible early twentieth century painters, both added a great deal of beauty and insight to the world through their art. And they also both said terri-ble things about each other. As theologian H. Richard Neibhur taught, we are more often right in what we affirm than in what we deny. Similarly, I see differ-ent schools of quality management all delivering quality, and being wrong when they say that other schools of quality management don't work.

These different schools are the subject of Part 3. In Chapter 11: *Total Quality Management* we look at the first comprehensive school of quality management, the father of all of the others. We will see how quality definition, quality plan-ning, quality control, quality assurance, and delivery of quality to the customer were integrated into one program for the first time. We will learn how compa-nies could adopt TQM, and see examples of their failures and successes.

There are three things that can define a movement and keep it together: per-sonality, standardization, and promotion through education, establishing professional standards, and marketing. Total Quality Management was unified by the personality of W. Edwards Deming and others, most particularly Joseph M. Juran. The newer movements—the descendants of TQM—have been held together by standards, or by marketing. In Chapter 11: *Quality Standards: ISO 9000 and More,* we look at how the International Organization for Standardization (ISO) has codified quality methods. This global standard is most influential in Europe. We will also look at the Malcolm Baldridge Award, which defines TQM in the United States, and at other influential standards. In Chapter 13, we will look at the most influential movement in North America, *Six Sigma.* First defined as a goal of manufacturing TQM, it has become a buzzword and a movement held together by marketing, training, and a common language. It focuses on some TQM processes, adding some refinements. It has had some spectacular successes, especially in reducing defects in manufacturing through applied statistics, and has also run into some challenges. In Chapter 13, you will learn the difference between successful and unsuccessful implementations of TQM.

In Chapter 14: *The Cost of Quality,* we will look at the work of Philip Crosby and others who introduced the goal of zero defects and the method of calculating the cost of quality. These developments were parallel to, but largely independent of, TQM. They also offer an important corrective and a tool for selling quality management improvement to executives. Using cost of quality analysis, we can define and implement quality improvement programs that are optimized to bring the biggest bottom-line bang for the bucks spent on improving quality.

Chapter 15: *The Capability for Quality: CMM and CMMI* traces the history of a standard that started with the goal of bringing TQM methods to the field of software engineering and succeeded in creating the world's first zero-defect software. The Capability Maturity Model (CMM) and its successor, the Capability Maturity Model Integration (CMMI) have become key standards for software development in North America, and are sometimes required for military and government contracts. In addition, industry in the People's Republic of China is rapidly adopting the model. Most importantly, the notions of maturity and capability give us language to define the largest challenge to improving quality in our companies—the nature of organizational culture, and how culture resists change to process.

We close Part 3 and finish our look at quality movements by returning to Japan. For over 50 years, Japanese industry and society have been steadily integrating and developing quality management. Starting with the development of TQM, they have integrated Crosby's work on Zero Defects, and developed innovations of their own. There is a lot to learn from a half-century of steady progress. In Chapter 16, we look at *Gemba Kaizen for Just In Time Manufacturing* (*JIT*), and see how quality management has taken over industrial management, creating a holistic approach that unifies production operations management, human resources management, cost management, and time management under one goal—delivering quality to maximize value. This approach integrates quality management into *lean manufacturing,* with the goal of customer delight through delivery of exactly the right quantity of products at the lowest possible cost through a focus first on quality, and then on efficiency.

Quality Management Demystified is, of course, too short to provide a thorough introduction to all of these movements. In the *Resources for Learning* section, you will find a list of excellent references, often written by or quoting the founders of these movements, who were lucid and clear speakers and writers, well worth reading.

Total Quality Management

> *[T]he fundamental principles of scientific management are applicable to all kinds of human activities, from our simplest individual acts to the work of our great corporations, which call for the most elaborate cooperation.... whenever these principles are correctly applied, results must follow which are truly astounding.*
> —Frederick Winslow Taylor, *Scientific Management*, 1911.

Total Quality Management (TQM) was the first method to successfully implement the ideas of Taylor and Shewhart more than one company at a time. It is the most central defining method in quality management. TQM is the father of all of the quality management methods and standards that have come since, including Six Sigma, CMM, ISO 9000, the Zero Defect movement, and, most directly, *Gemba Kaizen for Just-in-Time (lean) Manufacturing*.

Quality Management Before TQM

Before TQM, the field of quality management was not defined at the business level. Taylor's *Scientific Management* came closest. Although he was able to implement his methods himself, and he saw how others understood his management ideas, he was unable to pass along the ideas of executive leadership as a way of building followers and the importance of cooperation. As a result, once assembly lines were in place, process improvement was seen as a technical or, at best, management function, not something that would change the way companies were run or guided from the top.

It is interesting that Smeaton, Shewhart, and Deming were all physicists. Smeaton was able to convince engineers of the value of his ideas. And, in his time, that was enough. The proven results of his work gave him a place in history and transformed the field of civil engineering. Shewhart also transformed the field of engineering, but was unable to convince executive leadership in the United States of the value of his ideas after World War II. From their perspective, statistical quality control looked like an expensive solution to an even more expensive wartime problem related to inspection of products. North American executives did not see the business value—in reduction of waste and cost—of improving the quality of processes from the 1950s through the 1970s. Even though Shewhart's methods were essential to the new industry of the manufacture of electronics, leaders in his own country did not see how they could revolutionize all businesses. This is not surprising. In the post-war economic boom, American leaders had little reason to look for new methods. They had other priorities. They were preoccupied with increasing productivity, managing conflicts with workers and labor unions, and driving stock value by beating the predictions of industry pundits every quarter. And their jobs depended on dealing with these issues. Customers were a good deal less demanding than the army had been with regard to product defects—they appeared satisfied with product quality, even when it included built-in obsolescence. And Shewart was unable to explain how improvements in the quality of the process would improve productivity. As a result, Walter A. Shewhart spent much of his career from the 1950s on in other parts of the world.

One core problem of Fordism is that, on a practical day-to-day level, quality was always pitted against productivity, and productivity always won. Managers knew about quality problems and wanted to fix them, but pressure was always on to meet production quotas, and their jobs were measured by that. There was no reward for solving quality problems. At best, some time was allowed for QA when the factory wasn't busy. At worst, workers and managers were pressured to deliver defective products to meet quota, figuring that they could always be fixed later. And no one saw the fundamental relationship—demonstrated by Shewhart and Deming—that improving the quality of the process increased both productivity—quantity of product—and product quality at the same time.

The Core of TQM

Immediately after World War II, W. Edwards Deming attempted to teach American engineers the value of PDCA and statistical process control. The engineers learned it, but could not convince their management that it had value. Deming had helped the U.S. Census Bureau apply statistics to the national population census in 1940. As a result, the U.S. occupation government of Japan asked Deming's help in conducting the 1951 Japanese census.

In his first visit to Japan, Deming became an admirer of Japanese culture and also deeply concerned with the suffering created by the poverty of postwar Japan. Another group was equally concerned with the future of Japan—JUSE, the Union of Japanese Scientists and Engineers, the Japanese equivalent of the American Institute for Electrical and Electronic Engineers (IEEE). Members of JUSE heard Deming's first lectures on statistical quality control, and invited him back for more. Deming agreed, on the condition that the first class be open only to the leaders of the largest electronics and manufacturing firms in Japan. This condition was the result of what Deming had learned in the United States: engineers couldn't get the ear of executives. But maybe he could himself.

He did. Japan has a very hierarchical and cohesive culture, and it was in desperate need of transforming its industry. Japan is a small island without enough land to feed its modern population, so it must export something to buy food. But in the 1950s, Japanese exports were a joke. Materials management was so poor that no manufacturer could count on consistent raw materials or components, and there were no guidelines for controlling manufacture. *Made in Japan* meant low-quality goods. Japanese industry urgently needed to improve product quality to levels that would be acceptable in the world market. Deming told Japanese business leaders and engineers that, with quality management methods and hard work, Japanese products could compete on the world market within five years. They believed him, got to work, and succeeded.

One could say that Shewhart and Deming had a solution that fit the problems faced by business in Japan more clearly than it fit the problems faced by businesses in the United States. But this is not really true. It is easier to explain that quality in process improves quality in products solving Japan's business problem. It is harder to explain that quality in process prevents labor problems, increases productivity, and reduces cost. But both are equally true. In fact, the arguments for the value of quality management for the American situation were fully established in Taylor's 1911 treatise, *Scientific Management.*

It is essential to understand that TQM was created as a collaborative effort, combining the best of American scientific quality management with dedication, focus, and creativity provided by the Japanese people. Leaders of companies like SONY and Toyota made contributions. For example, here is part of the vision statement written by SONY's founder, Masaru Ibuka:

> If it were possible to establish conditions where persons could become united with a firm spirit of teamwork and exercise to their heart's desire their technological capacity—then such an organization could bring untold pleasure and untold benefits—Those of like minds have naturally come together to embark on these ideals.

Purpose of Incorporation

- To establish a place of work where engineers can feel the joy of technological innovation, be aware of their mission to society, and work to their heart's content.
- To pursue dynamic activities in technology and production for the reconstruction of Japan and the elevation of the national culture.
- To apply advanced technology to the life of the general public.

Management Guidelines

- We shall eliminate any unfair profit-seeking, persistently emphasize substantial and essential work, and not merely pursue growth.
- We shall welcome technical difficulties and focus on highly sophisticated technical products that have great usefulness in society, regardless of the quantity involved.
- We shall place our main emphasis on ability, performance, and personal character so that each individual can show the best in ability and skill.

Excerpted and translated in *Built to Last: Successful Habits of Visionary Companies*, by James C. Collins and Jerry I. Porras, HarperCollins, 1994.

It is clear that Japanese executives contributed to and embodied the spirit and leadership potential of quality management. Equally important, quality management became a company-wide team effort at all levels, including engineers, technical teams, managers, foremen, and workers. We will learn more about this in Chapter 16: *Steady Improvement in Japan.*

The team element of early Japanese TQM was perhaps an accident, but it certainly was a success. Deming could only train groups of about 50 engineers at a time, and his trips to Japan around 1950 were sporadic. Only a few people from each company were able to attend and receive the training. They brought back the methods and launched study groups. As soon as the study groups understood the basic ideas, they got to work implementing them. The economic conditions in postwar Japan certainly added to their drive for improvement and success. The study groups became departmental quality teams and cross-functional quality teams, core elements of TQM. And—in less than five years—the Japanese were exporting high-quality manufactured goods all over the world.

FROM JAPAN TO THE WORLD

Japan is a small country on the world market, and it took 20 years for its success to make a really big difference in U.S. markets. I remember, as a child in the late 1960s, wondering why everyone thought *Made in Japan* was a joke while we were snapping up SONY transistor radios and TVs. By 1980, when I was in college, we learned that American industry was in trouble, and Japanese management methods might be the solution. In the 30 years from 1950 to 1980, American business rode the postwar economic boom like a wave, and, even when the wave petered out, our momentum kept us rolling. Increasing productivity to meet market demand, growing market demand through advertising, and keeping the unions happy were the central focus of management. We did not see the troubles created by our own internal processes, even while these troubles were beginning to erode customer satisfaction and global market share, made people doubt our advertisements, and were part of what sustained the conflicts between management and workers.

THE PROBLEM THAT TQM SOLVED

By the late 1970s, U.S. industry knew it was in trouble. In terms of market share, the problem was deeper than loss of ability to compete in world markets. Our own companies began to lose ground in our own country. Japanese makers of photocopiers were so cost-efficient that they could produce copiers in Japan, ship them to the United States, and sell them at a retail price lower than the manufacturing cost of a functionally identical machine made by Xerox Corporation here in the United States. At first, we thought lower foreign prices were due to lower foreign wages or the fact that Japanese and German manufacturing infrastructure were newer—and therefore cost less to operate—than America's aging factories. But the real cost saving came from Japan's *kaizen*—continuous improvement of processes to increase quality while reducing cost. U.S. industry and the American popular media were looking for a solution.

In 1980, they found one in their own backyard. W. Edwards Deming, an American and a Japanese National Treasure, was discovered living just two miles from the nation's capital. Symbolically, the pivotal moment was when the CEO of Ford Motor Company called W. Edwards Deming, then 80 years old, and asked him to come in and guide them in fixing their company. That began a vigorous period of progress in quality management in North America and around the world that lasted more than a decade. In that time, the methods developed by Deming and the Japanese were empirically proven to transcend cultures. The term Total Quality Management (TQM) was coined, and many companies adopted programs—typically five years long—to transform operations. The first goal was to be able to deliver high quality reliably. The second was to reduce cost. The reason was to stay in business.

One story illustrates that what TQM offered was a fundamentally different approach to management, and not simply quality methods. Around 1980, a U.S. television manufacturer could no longer afford to keep its factory and assembly lines running. They planned to close the business, sell the machinery and the land, and lay off the workers. A Japanese electronics company came and offered to buy the whole plant with equipment, use it to make televisions, and keep as many of the workers as they could. The deal was too sweet to pass up, and the American executives didn't ask too many questions until after the papers were signed. Then one American executive inquired: "We're very happy to see this deal go through. But I've been wondering—given we couldn't make televisions profitably in this plant, how is it that you plan to do it?"

A Japanese executive replied, "Right now, you have a 5% failure rate in the final quality assurance check before shipping. We plan to reduce that to 0.2%. Then we won't fix those two TVs per thousand, we'll just throw them away. With no rework line, the company will be profitable."

The American executive, a financial officer, replied, "It's true that our entire profit is eaten up by the rework line. If you can do it, it will work. But 5% is the lowest failure rate in the electronics industry!"

To which the Japanese executive politely replied, "not in Japan."

Using the same equipment and the same workers, Japanese executives made the plant profitable by eliminating the cost of poor quality. What is amazing is that they did it with just two changes in the management approach:

- *Listening to workers and using their expertise.* Fordism kept suggestion boxes in place to keep workers happy, but never paid attention to the suggestions. As one Xerox executive ruefully put it after the company's transformation by TQM: It is hard to admit now, but in the old days, when a worker came to us with a suggestion, we would tell him, "We hired you for your hands, not your head." The Japanese managers took the opposite approach. They went to each worker and said, "This piece of the assembly line is your world all day long, and you know more about it than anyone else. How would you redesign it to make it better for you, and so you could do better work?" Then they implemented the suggestions before restarting the assembly line under the new ownership.

- *Fix the process, not the product.* When the final QA/QC was being done at the end of the assembly line, a new approach was taken when a defect was found. Instead of fixing the TV, the new management fixed the process. Each time a defect was discovered, the assembly line was halted. The inspector would go to the station where the error had been created, and offer support to change the process so that the problem would never happen again. Once the permanent preventative solution was in place, the assembly line was restarted with a new process that was free of one more error.

Using just these two management methods, the new management—in cooperation with the old workers—was able to reduce rework from 50 TVs per thousand down to 2 TVs per thousand—two sigma quality to three sigma quality. TQM had arrived in the U.S.

Essentials of Total Quality Management

This is what the "total" means in Total Quality Management:

- *Everyone focuses on quality* from senior management all the way to the workers.
- *Everyone is trained in PDCA and other TQM techniques.* Once empowered and led by Theory Y managers, workers—if Theory Y is right—will enjoy their work and gain satisfaction from improving quality and delivering results. Managers focus on removing roadblocks to quality and productivity.
- *Everyone is empowered to make change.* Ideas are solicited from everyone, and, as much as possible, process improvement is carried down to the management front line and to the workers.
- *Departments become self-managed teams* focused on quality and productivity.
- *Cross-functional teams* solve problems all through the assembly line or workflow.
- *When the company finds a problem outside, they collaborate to fix it.* TQM is extended up the supply chain to raw materials and downstream—through

customer delight—through marketing, sales, customer service and support, to the smile on the customer's face.

- *Using fact-based decision making,* executives set direction and managers and workers improve process based on the evidence drawn from observation and experiments. Ideas, instead of being political footballs that only move ahead through personal persuasion, are weighed and tested. In this environment, people become more willing to present alternative solutions, knowing that, if they are shown to be best, they will be implemented.
- *Quality becomes an internal value.* In an environment that supports change, some people take on TQM methods for professional growth or transformation of character to develop leadership abilities.

Here is how a company transforms itself with a TQM implementation:

- *From the top down.* A TQM program begins with training for executive management. These executives commit to the program and then lead and influence others as they are trained.
- *A program to change processes.* Deming stressed that conversion to TQM would take a large manufacturing company five years. In those five years, the goal was not just to reach certain quality levels, but to change thinking about productivity and quality, learning to create a company that could consistently deliver high quality at high volume.
- *Continuous improvement.* TQM methods, once learned, can always be applied again. PDCA is the core of an iterative cycle of continuous improvement that leads to better and better processes—removing bottlenecks, eliminating sources of customer dissatisfaction, and continuously making change to be more efficient and more effective.
- *Capability to handle changing markets.* TQM's reduced cycle time, focus on rapid transfer of knowledge, and focus on customer delight allows companies to maintain a competitive edge. The company can see changing preferences, changing competition, and changing technology, seize on opportunities, and implement new solutions rapidly.

Deming's 14 Points—A Framework for Quality Management

I'm going to make a radical proposition: I believe that, in terms of workable ideas, TQM is a conceptually complete framework that can be used to evaluate any quality management method or program. I mean that any quality management improvement program, initiative, or ongoing process—no matter what it is called—will succeed if TQM principles are followed and well implemented, and will run into trouble to the extent that they are not. We will evaluate this proposition in this part of *Quality Management Demystified.* I will present the core principles of TQM in this section. Then, as we evaluate the strengths and weaknesses of each method and movement, we can see if the strengths come as a result of adhering to these 14 points, and difficulties arise when they are not given sufficient attention.

I am going to be liberal in revising some of the language of Deming's 14 points. Each of the 14 points expresses an essential principle, but Deming sometimes expressed this principle in the context of specific business issues. If we do not understand the assumptions of the time and the people he was dealing with, the points can seem obscure. The points will be presented as clear principles, and then I will explain how I derived them from Deming's original material. When the point appears in quotation marks, it uses Deming's own words. In two cases, I made mild revisions to the name of the point and then explained and supported my choice with quotes in the same paragraph. All quotes in this section are from W. Edwards Deming. This section draws from Mary Walton's *The Deming Management Method.* Mary Walton was instrumental in the U.S. discovery of Deming around 1980, and he authorized the preparation of Walton's book and assisted with the manuscript.

1. *"Create constancy of purpose for the improvement of product and service."* A company needs to focus on the customer and long term success, and not "stay bound up in the tangled knots of the problems of today, becoming ever more efficient in them." There should be leadership through support of innovation and education, continuous improvement of product and service, and investment in necessary equipment.

2. *"Adopt the new philosophy."* Deming says, "Point two really means in my mind a transformation of management. Structures have been put in place in management that will have to be dismantled. … We will have to undergo total demolition of American-style management, which unfortunately has spread to just about the whole western world." This is essential because the old approach is full of hidden costs that ultimately reach the customer, making the company not competitive. In the sphere of society and government, the same is true, and the taxpayer pays.

3. *Use the results of inspection to design in quality, not just to find and remove failed products.* This point, originally phrased as "cease dependence on mass inspection," is one that has been badly misunderstood. As we addressed in Part 2 and in the story of the TV assembly line, the issue is not inspection, but rather what we do with the results of inspection. Deming says, "Inspection with the aim of finding the bad ones and

throwing them out is too late, ineffective, costly. In the first place, you can't find the bad ones, not all of them. Second, it costs too much. The old way: Inspect bad quality out. The new way: Build good quality in."

4. *Choose suppliers for commitment and ability to deliver quality at a good price, not just on price alone.* Originally, this was phrased as, "End the practice of awarding business on price tag alone." Deming points out that when a company uses only short-term contracts and jumps from one supplier to another, lower quality materials and components is the inevitable results. Suppliers competing on cost can't afford to customize to meet your needs without knowing they will continue to be your supplier. Deming recommends working with a single supplier using a teamwork model with both committed to collaborating for high quality and cost management over the long term.

5. *"Improve constantly and forever the system of production and service."* The point here is that even keeping a process under statistical process control by removing special causes of variance is not improvement. "Putting out fires is not improvement. Finding a point out of control, finding the special cause and removing it, is only putting the process back to where it was in the first place. It is not improvement of the process." Deming attributes this conclusion to Dr. Joseph M. Juran. Juran's attention to radical improvements in method was an inspiration for the Six Sigma quality management movement.

6. *"Institute training and retraining."* Here, Deming echoes Taylor on management's responsibility to define best practices and then train workers, and also to work with workers to improve practices and bring everyone up to the level of newer, better management. He says that every worker should have a basic understanding of control charts. His thinking foreshadows ideas further developed in CMM about the nature of skills transfer being an essential element of a company's capability to produce consistent high quality.

7. *"Institute leadership."* Deming holds that almost all poor performance is a result of poor management, where poor supervision has replaced leadership and establishment of quotas has replaced the setting up of effective systems, the training of workers to be self-managed, and a focus on removing barriers to departmental success. Of course, some workers will be simply unsuited for a job or unwilling to work, and management must deal with that. But these cases are the rare exception, not the rule. Deming believed that 80 to 95 percent of all productivity problems were the responsibility of management, because a system under statistical process control eliminates that percentage of problems, and putting controls in place is the job of management.

8. *"Drive out fear."* Deming said that feeling secure (which in its Latin roots, means "without fear") is essential for effective communication about the job and about problems. Problems cannot be fixed when people are afraid to speak up, whether it is fear of losing a job, losing an argument, or being blamed for a problem. Deming's concerns about fear were even wider: "The economic loss from fear is appalling." "Fear takes a horrible toll. Fear is all around, robbing people of their pride, hurting them, robbing them of a chance to contribute to the company. It is unbelievable what happens when you unloose fear." I suggest that we recognize that this principle goes deeper than just eliminating blame and anger from the workplace and ensuring good workers a stable job. It also means influencing social factors to reduce fear and increase genuine security through cooperation and trust. I also believe that this one point, more than any other, is what is missing in many failed quality programs, no matter what method they use.

9. *"Break down barriers between staff areas."* This is one of Deming's points that has been accepted perhaps more than many others. At least almost all businesses now recognize the problem, often called a stovepipe mentality. Deming asserts that cross-functional teams and efforts to solve problems throughout what we now call the SIPOC chain, combined with good statistical quality control in each area, lead naturally to lean or just-in-time manufacturing. Once the barriers are broken down, communication and problem solving open the doors to working out the kinks—to streamlining stable processes. But going for just-in-time is "sheer nonsense unless the process is stable. Unless it is stable, nobody knows who is going to need what or when he'll need it."

10. *"Eliminate slogans, exhortations, and targets for the workforce."* Slogans, he says "generate frustration and resentment." Instead, change the system. The only point I would add is that anything can become a slogan, and almost anything can become a genuine goal and direction. "Zero defects," as a slogan, doesn't work. The zero defects movement is the right solution in certain work environments when it is properly implemented. Similarly, any of Deming's 14 points on this list could become a slogan, if management didn't make the necessary changes to support the team in achieving the goal.

11. *"Eliminate numerical quotas."* This would better be stated, "eliminate purely numerical quotas." Deming was more concerned about the absence of quality requirements than he was with the presence of the numerical requirement. Deming indicates that when we eliminate numerical quotas, we should replace them with a quality requirement and accepted levels of quantity with variation. Allowing variation prevents tampering, as we

discussed in Chapter 10: *Statistics for Quality.* Then we use PDCA and other TQM techniques to improve the process to maintain quality and increase productivity. This approach also stresses cooperating towards solutions instead of demanding that fires be put out. Deming stresses that quotas are just as inappropriate for managers as they are for workers.

12. *"Remove barriers to pride of workmanship."* Poor management leading to unpredictable changes of work and requirements, requests or demands to do inferior work, and failure to provide solutions to problems with tools, inputs, and the work environment all quell the employee's natural desire to do quality work and deliver results. We replace these problems with communications about problems and prompt solutions to problems—empowering workers to speak up, to do good work—and supporting them by solving problems and removing roadblocks that are not under their control.

13. *"Institute a vigorous program of education and retraining."* Here, Deming makes four points: The company will change as a result of a quality improvement program, and some job slots may be lost, but no one should lose their job due to increases in productivity. Instead, they should receive training for new positions. Secondly, everyone should learn the basics of PDCA and statistical quality control. Third, retraining should be a specific investment in people's ability to do their changing jobs. Fourth, any and all general education is of tremendous value. It should be pursued by the employees and supported by the company. "You never know what could be used, what could be needed. He that thinks he has to be practical is not going to be here very long." And, to businesses, "Help people to improve. I mean everybody."

14. *"Take action to accomplish transformation."* On this subject, Deming stressed everyone learning and using PDCA; everyone seeing that their work has an internal or an external customer and seeking to satisfy that customer; and everything being seen as a process that can be improved.

Is TQM a Total Solution?

Clearly, if management in the 1990s had seen TQM as a total solution, we would still be using it, and we would still call what we do TQM. Instead, in the 1990s, companies either got frustrated with TQM, or faced new problems and thought that TQM couldn't solve them. I would suggest that this reflects more of a failure by the company to fully engage points one and two, rather than a failure of

TQM. The pressure to show quarterly profits, the allure of quick fixes and new approaches, and the fact that deeply ingrained attitudes are hard to see and change all contributed to companies falling away from TQM.

One of the key issues here is that processes can only change if culture changes, and culture can only change if people are willing to change. For people—especially senior executives—to change, there needs to be a good reason. For most people, a crisis is sometimes that reason, and sometimes a reason that fades in a few years, so that the importance of the change and the commitment to it fades as well.

TQM companies did face challenges, but I believe they could have been resolved by applying the core principles of TQM, even when that meant breaking some other rules laid down by Deming in an earlier era. For example, when Deming said that it takes five years of total commitment to implement TQM, he was referring to large manufacturing firms. Also, he was talking about current best practices. Companies in the 1990s did need faster methods of improvement. Rather than turning away from TQM principles and using process re-engineering that included cutting jobs (violating TQM point #13), the companies could have used best practices developed within TQM and improved while weathering the storm. Faster solutions, instead of replacing TQM, could have worked within the TQM framework. And some companies did just that. For example, SONY USA still operates with TQM. Motorola and GE evolved from TQM to Six Sigma without slipping off the quality bandwagon. Other companies did slip off and, in resorting to downsizing, allowed fear to re-enter the workplace.

As we continue our look into more recent methodologies, we will see if anything was missing from TQM; if any of the newer methodologies have added anything substantially new, or if they have simply repackaged TQM principles with a slightly different spin.

Conclusion: TQM—First Among Many

Total Quality Management was the first quality management method to transform entire industries, rather than work at just one company at a time. In the following chapters, we will learn about what has come since, and see how other methods critique TQM, and how they shape up against it, and what they have to offer in and of themselves. First, we will turn our attention to the standards, including ISO 9000 and the Baldridge Quality Award, that seek to keep everyone on the same page and move forward together when it comes to quality management.

Q-Ball Quiz for Chapter 11

1. A key problem with Fordism is that
   (a) it only works for American companies.
   (b) it pits productivity against quality.
   (c) it only works for manufacturing companies with assembly lines.
   (d) it strives to reach Six Sigma quality before putting basic procedures in place.

2. Which of these is the most accurate description of how TQM developed?
   (a) Japanese industry followed a plan laid out by Deming.
   (b) Japanese industry and Deming laid out an improvement plan and followed it.
   (c) Japanese industrial leaders, Deming, and managers and workers in Japan at all levels contributed to the creation of TQM.
   (d) It took only five years for TQM to develop fully in Japan.

3. Which of these is true of Deming's 14 points?
   (a) They focus on management, not leadership.
   (b) They emphasize quality, saying it is more important than productivity.
   (c) They include education, elimination of fear, and restoration of a worker's pride in his work.
   (d) They offer a step-by-step process for transforming a company.

4. Which of these is *not* one of Deming's 14 points?
   (a) Drive out fear.
   (b) Maintain stable production with quality control.
   (c) Break down barriers between staff areas.
   (d) Institute leadership.

5. Which of these guidelines for transforming a company through a TQM program would not always apply?
   (a) It will take five years.
   (b) It must be done from the top down.
   (c) Everyone must receive training.
   (d) Communication across departments must improve.

Quality Standards— ISO 9000 and More

The general adoption of scientific management would readily in the future double the productivity of the average man engaged in industrial work. Think of what this means to the whole country. Think of the increase, both in the necessities and luxuries of life, which becomes available for the whole country, of the possibility of shortening the hours of labor when this is desirable, and of the increased opportunities for education, culture, and recreation which this implies. But while the whole world would profit by this increase in production, the manufacturer and the workman will be far more interested in the especial local gain that comes to them and to the people immediately around them.
—Frederick Winslow Taylor, *Scientific Management*, 1911.

It is one thing to do quality work for your customers; it is something else to be able to prove it. But do we need to prove it? If quality is really valuable to companies, then won't our quality management system prove itself in customer delight leading to higher revenue and greater market share, while reducing waste and cost?

Those are certainly the most important places to see the results of a quality management system. But there can be added value in adopting external standards and receiving certification:

- *Getting a quality management system as a package may be cheaper than reinventing the wheel.* If you need a fresh start, why not hire experts, get everyone trained, adapt the system to meet your company's needs, and then, as a bonus, get registered or certified, as well?
- *Sometimes, registration or certification is a customer requirement.* In some cases, our customer might request or require that we, as a vendor, meet the requirements of external standards and be able to prove it. Certification is the more common term for this kind of application for third-party approval, but the ISO uses the term registration. In *Quality Management Demystified,* we will use the terms interchangeably.
- *An expert opinion can show us what we can't see about ourselves.* Even when our car is running fine, it makes sense to take it in for a tuneup. Similarly, we may think that everyone is doing a good job, but review by an auditor may show us surprises in time to fix them. External standards can be adopted as internal guidelines, or as a framework for internal guidelines. When we do this, we can make our own quality management system more effective for ourselves and our customers, and have that validated by an external registrar.

Of course, it only makes sense to adopt a standard if the standard is either required, or it actually improves quality and performance. We'll take a look at ISO 9000—the most significant worldwide quality standard—and others, so you can see for yourself what standards can do for your company.

ISO 9000

The ISO, or International Organization for Standardization, is a worldwide body that sets standards for industry. Until the 1980s, most of the standards were technical. For example, the ISO defined the OSI (Open Systems Interconnect) model that is used for both the worldwide telephone network (implemented as Signaling System 7) and the Internet (implemented as TCP/IP). Such standards are beneficial to business because they allow integration and innovation. For example, many companies can build telephone equipment, and they all work together around the world as a result of the SS7 standards. Because SS7 and

TCP/IP are both based on the OSI model, they can be integrated with the new technology called VOIP (Voice Over IP), or Internet Telephony.

Although global in scope, the ISO is centered in Europe, and its standardization has been a force supporting the gradual economic and political unification of European countries. In particular, the ISO 9000 series of quality standards makes international commerce with European partners easier.

Variations on the ISO 9000 standard have also been adapted by U.S. auto makers and by the aerospace industry. In each case, the goal is to do with quality management—a business process—what earlier ISO standards did with technical interfaces. The idea is that, if each company complies with the same standard, then it will be much easier for any one company's products to be input in the SIPOC chain for the next company. Rather than specifying specific technical interfaces, ISO requires companies to create their own quality management system (called a quality system or QS). There is a great deal of flexibility in the design of your QS, but it must be focused on meeting customer requirements. The idea is that if the customer side of any interaction has clear requirements, and each company operates to the ISO 9000 standard, industries will be able to design and deliver complex systems—such as cars, elevators, or airplanes—that are safe and of high quality.

WHY ISO 9000 MATTERS

To comply with the ISO 9000 standard, a company must set up its own internal Quality System that basically ensures three things:

- A focus on quality defined as customer satisfaction and meeting customer requirements.
- That we, as a company, do what we say and say what we do, and are able to prove it.
- That we use PDCA for correction of problems.

The benefits of actually doing these three things are already clear. A company meets ISO 9000 standards by doing what Taylor demonstrated was the best way of doing business back in 1911. But what is the significance of registration—the process of becoming certified as in compliance with ISO 9000 and maintaining that certification?

First of all, your customers may demand it. If you have corporate customers in Europe, this is increasingly likely. But it can happen in the United States as well. There are a number of industry-specific standards similar to ISO 9000 in various stages of development. The aerospace standard is AS 9000. The big three automakers in the Unted States. created QS 9000, but that is in a state of fluctuation, with another standard, ISO/TS 16949 coming into play. The variations are industry-specific and rapidly changing, so there is no point detailing them in *Quality Management Demystified.* Instead, we will simply use the term ISO 9000 generically to refer to all its variations and similar standards. Two other trends are worth noting. One is that some government agencies may require ISO 9000-type certifications of their vendors. The other is that, after major companies require their direct vendors—called Tier 1 vendors—to be certified, those companies may then require their own vendors—called Tier 2 vendors—to be certified as well. This is important because it can mean that rather small companies are caught by surprise and suddenly have to rush to get registered or certified, which can be a costly and painful process. It is also relevant for companies changing the market they serve, as ISO 9000 compliance may be an entry requirement for certain markets.

Q-PRO

Why a Standard?
This story explains how and why customer quality requirements became industry quality standards in the U.S. auto industry.

The creation of QS 9000 by the big three U.S. automotive companies—GM, Ford, and Chrysler—is something from which we can learn. Each company had found that, in order to meet their own quality and productivity goals, they required vendors to meet quality management requirements as well. That is, they imposed requirements on their supply chain. Initially, each of them had their own proprietary requirements, and also maintained expensive external audit functions that would audit their vendors and suppliers to ensure compliance. All of this was very expensive and each of the big three was footing its own bill.

Inspired and informed by ISO 9000, they got together and adopted ISO 9000 with some automotive-industry specifics and additions, and called it QS 9000 as an independent standard to be audited by a third party. This reduced the costs for the big three. The benefit for the vendors was that they could meet one standard and serve three customers. The cost for vendors was that they had to pay for their own audits to get registered and stay registered.

Note that QS 9000 now governs the vendors' process. Customers still define the product specifications; but since the U.S. auto industry embraced TQM in the 1980s, the car makers saw that they needed to be assured of their vendors' process as well. So that has now moved—from being a two-party relationship between vendor and customer to being a three-party relationship maintained through certification to an external standard.

The lesson: Standardization reduces the total cost of quality in an industry by allowing companies to meet one standard and serve multiple customers. However, it may also shift those costs from one party to another.

THE ELEMENTS OF ISO 9000

ISO 9000 requires only that a company be able to demonstrate that it embraces the principles of quality management and has a quality system (QS) in place to make sure that it is actually doing so. Each company defines its own specific goals and methods. For example, one company might have a practice of eliciting detailed customer requirements for custom work, while another does general market studies to define generic products. One company might choose inspection, another statistical quality control. One company might value innovation, another efficient service.

What is required is that your company define all of these goals in writing, develop written processes for all work related to these goals, and put a quality management system in place that assures that these goals are being met and that work is actually being done as specified. Specifically:

- We create a quality mission statement and a quality manual.
- We define documents that describe how work is to be done. ISO 9000 refers to these as documents; a common term is standard operating procedures (SOP).
- We define a quality system that can produce reports showing that the SOPs were followed in doing the work. Reports would include checklists and records of QC and QA data.
- We train everyone in the parts of ISO 9000 relevant for their work. This is usually a one-day training. Costs can be reduced if workers are trained to write, or at least modify, their own SOPs.
- We set up a QC system that uses PDCA to correct any nonconformities discovered by our quality system.
- We set up a QA system that uses PDCA to ensure that SOPs are correct and up to date, and that ensures that issues discovered about our process move into quality improvement efforts.
- We set up an internal quality auditing system that ensures that all of the above is being used and maintained. We do not need to have fulltime internal quality auditors; we can use staff with other duties part time. But we must have enough internal staff trained in QA that there is someone available to audit any work who was not involved in that work. The primary difference between quality assurance and quality auditing in ISO 9000 is the degree of independence of the auditor.

The ISO 9000 series of standards is divided into sections. ISO 9001 is the most comprehensive. Higher numbers—ISO 9002, 9003, and so forth—are standards focused on particular functions or industries, or more detailed guidance in how to meet these standards. Some of these are options for registration, others are for your information. Recently, ISO has added implementation guidelines in ISO 10000, but these are generally considered together with the ISO 9000 series.

HOW TO GET ISO 9000 CERTIFIED

The least expensive and fastest way to prepare for certification is with the help of outside expertise. ISO 9000 consulting and training firms can provide you with complete training for all staff levels, build a plan with you, and provide you with templates for all required documents and reporting systems.

Meeting ISO 9000 requirements is like doing the first year of Deming's five-year approach to transformation through TQM. Key elements are top-down commitment, company-wide training, and employment of PDCA first to bring production under control of inspection or statistical analysis, and then to improve any and all parts of our organization. Like TQM, ISO 9000 only succeeds if it is integrated with organizational management and resolves the conflict between productivity and product quality.

Here are the specific steps to prepare for ISO 9000 registration:

- Evaluate to see if ISO 9000 is right for your company. If it is, make an executive commitment to doing it and engaging the full support of your team.
- Name an executive representative who will communicate with ISO external organizations and lead the internal quality council made up of senior managers and managers.
- Establish one or more quality management implementation teams.
- Perform an internal study to establish a baseline—a report on the current situation of your quality system—and to define the gaps that need to be closed so that your QS meets ISO 9000 requirements. Expert consulting is useful for this, as you do not want merely to meet your own interpretation of ISO 9000 requirements.
- Build an implementation plan that closes all the gaps between your current QS and ISO 9000 requirements.
- Document everything to show evidence of thorough planning and to demonstrate that every process that affects quality has been thought through, planned, and turned into a document or SOP to be followed. The document to be followed can be called a facility quality manual. This is essential for the auditors.
- Maintain your new quality system for three to six months, including auditing it and correcting any nonconformities.

At this point, you are ready to seek formal registration. You contact an ISO 9000 registrar. A *registrar* is an independent company that, for a fee, audits you to see if you meet—and later to see if you are maintaining—compliance with the ISO 9000 standards. Each registrar is, in turn, registered with an agency—a different one in each country around the world—that ensures all registrars are auditing independently and properly ensuring that they only issue certificates to companies that truly meet the ISO 9000 standard. The process of completing registration looks like this:

- Select a registrar you want to work with.
- Apply for registration.
- The registrar performs a document assessment. In traditional audit language, this is called audit planning. All of your documents are reviewed. The primary question in mind is: Do these documents establish controls that should be in place for all functions that affect quality throughout the organization? If not, you will be asked to remedy these nonconformities before the audit.
- Once controls are demonstrated to exist in the documentation, the registrar sends an audit team to investigate if the controls are actually in place and being used. This includes interviewing personnel to determine that they know and follow correct procedures, and know what to do if a nonconformity is found in a product, or if a procedure needs to be updated. This is the part of the process where you prove that you do what you say you do.
- If any major nonconformities are found, you remedy them, and the registrar checks them in a brief followup audit. In some cases, minor nonconformities can be recognized without delaying certification. You will merely need to demonstrate compliance in time for your first maintenance audit.
- You pay your fee and receive your certificate.
- You maintain your IS system in use and continue your internal audit function. Typically, internal audit should try to check the entire organization once a year, but this is not an ISO requirement.
- Every six months, you arrange for a registrar to perform your external audit so you can maintain registration and keep a valid certificate.

ISO 9000 originally was primarily focused on manufacturing, especially in fields where human safety was a major product concern. This included companies that produced elevators, components for cars and airplanes, and so forth. To some extent, general manufacturing in fields that use less sophisticated engineering has adopted ISO 9000 as well. In addition, use of ISO 9000 can also be applied to service industries. Some say that ISO 9000 still has an engineering/manufacturing flavor. Companies in software development might prefer CMMI to ISO 9000, as it is industry specific. Also, in choosing a consulting firm or registrar, it makes sense to pick someone who understands your industry and someone who you believe will work well with you.

This last point raises an interesting issue. Is ISO 9000 implemented by different registrars with essentially the same rigor? Or does a company get off more lightly by choosing a more lenient registrar? I was discussing ISO 9000 and CMMI with the head of quality for a major research firm, and he said simply, "There is a lot of shopping around for registrars." But the question is, who is really better off? Is it the company that chooses a more lenient registrar and has to do less work to come into compliance, or the company that chooses a more demanding registrar, and is pushed to achieve higher quality?

PRACTICAL PREPARATION AND MANAGEMENT

Sometimes, ISO 9000 is criticized as a mountain of paperwork. The reality is that it is the minimum of paperwork necessary to maintain a QS. But companies that do not have standard practices in place will need to organize a lot of information. There are two crucial issues: documentation and record management; and isolation of nonconforming products.

The ISO 9000 documents—manuals and SOPs—are living documents that can change frequently as you improve your business processes. You need a way to keep track of them all and make sure everyone is on the same page and using only the latest version. This could be implemented through any number of relatively simple or more complicated software document management systems. Of course, there are many more records than there are standards documents. Records of every process completed—usually checklists—and every test result, both human and machine, need proper storage. Even more, we need systems that highlight nonconformities in process or product and bring them to our attention. And all of this needs to be organized in such a way as to leave an audit trail.

The other issue is the isolation of nonconformities. This means two things: First, if a process isn't working, you stop it. Second, if something fails a check or test, you pull it off the line and don't send it to a customer. Both of these sound obvious and simple but, in actual practice, they are not. There are several things that keep us from handling nonconformities well, even when we know about them:

- *Human error, inattention, and interruption.* We simply make a mistake, sending a bad item instead of a good one, or miss a sign that something is wrong, or get interrupted and don't get back to fix what we were trying to fix.

- *Pressure to produce.* There are many stories of workers being directly told to produce defective parts, and cases of items known to be defective being shipped.
- *Lack of connection of process to product.* How often have you eaten something that was not well cooked? The process was bad, but you tolerated the product, usually to avoid waste or out of an unwillingness to confront the person who did a bad job. Is your company doing that to its customers.
- *Lack of space.* If we haven't planned for handling of defective products, we may not know what to do with them and so may not have a good way to get them out of the system. They might be stored to one side, but then loaded on the customer's truck by accident.

To meet ISO 9000 requirements, you will need to manage your records and your products well.

PLUSES AND MINUSES OF ISO 9000

As a quality management methodology, how does ISO 9000 stack up against TQM and other methods? It is rudimentary. To put it another way, it is a first step, but it is the right first step. A company that truly wants to transform through quality management would do well to begin with ISO 9000. I mean this in two senses. First, adopting ISO 9000 will be the equivalent of the first year of a TQM initiative. Second, adopting ISO 9000 requires a company to establish clear quality requirements and procedures, which is an essential step in preparation for the application of statistics to quality problems.

In one way, ISO 9000 does move in a direction that is different from TQM. The emphasis on auditing—both internal and external—is greater. On one level, this simply reflects the fact that, in going for ISO 9000 registration, a company is setting out to prove that it has an effective quality system in place, rather than just allowing its products and customer service to demonstrate that to customers in the marketplace. On another level, it could reflect a deeper issue of corporate culture. In TQM, the way a culture of quality spreads through a company is through the influence and guidance of leadership. This is supported by high levels of training and by recruitment and indoctrination practices that select people who fit in well with corporate culture and values. As long as people who are committed to the company and to quality are selected, they can be included in the TQM company somewhere, and can move to a job that is a good fit if necessary. This fits very well with Theory Y management.

ISO 9000 is possibly more compatible with Theory X management. It relies more on independent validation proving that good work is being done, and less on influencing each worker to evoke a commitment to quality. It puts more resources into independent checking, and less into training. ISO 9000 can certainly be adapted to a TQM company at reasonable cost, because any TQM quality engineer can easily qualify to be an ISO 9000 quality auditor, and then offer services to other departments. But, if a company is in a situation where, due to culture or other factors, it cannot influence its entire company to accept TQM, then ISO 9000 might actually be preferable. For example, if a company uses consultants or brings in new employees for short-term project work, then ensuring quality through the ISO 9000 audit approach might be more effective than providing a lot of TQM training to people who will only be with the company for a relatively short time. Or if a company is organized as a loose cluster of acquired companies, it might be easier for the highest executive level to require ISO 9000 and support other initiatives, but allow the relatively independent operating groups leeway in choosing when and how to move quality beyond ISO 9000 to a total approach.

Other Awards, Standards, and Associations

THE BALDRIDGE NATIONAL QUALITY PROGRAM AND AWARD

The Baldridge Award was established by the U.S. federal government in 1987, during the heyday of the American TQM era. It is administered by the National Institute of Standards and Technology (NIST), a division of the United States Chamber of Commerce, supported by the American Society for Quality (ASQ). It was named after the U.S. Secretary of Commerce from 1981 to 1987, Malcom Baldridge. Winning a Baldridge Award was considered a demonstration of the highest achievement in TQM, and many companies sought the distinction. Designed to be similar to Japan's Deming award, it has evolved over the years. Originally focused on manufacturing, it now focuses on the organizational capability for quality and specifically gives awards in five areas: manufacturing, service, small business, education, and healthcare. Two interesting aspects of the award are that it applies to not-for-profit as well as for-profit organizations, and that organizations traditionally not considered to be part of business—such as school districts—can win the award.

Applicants are evaluated in the following areas:

- Leadership
- Strategic planning
- Customer and market focus
- Measurement, analysis, and knowledge management
- Human resource management
- Process management
- Business results

In 2005, there were 64 applicants for the award: 33 in healthcare; 16 in education; 8 in small business; 6 in service, and only one in manufacturing. Full information and applications are available at *www.baldridge.nist.gov*.

Evaluation of the Baldridge Award

The Baldridge Award requirements are established from a customer-oriented, results-focused TQM approach. Achieving a Baldridge Award requires a great deal more effort than achieving ISO 9000 registration. And the award is a one-time event. Unlike registration or certification, it is not supervised for ongoing compliance. In fact, award recipients are ineligible for five years after winning an award.

One of the most interesting ideas about the award's benefits is that applying for the award is a very inexpensive way to get a lot of topnotch quality management consulting. Each applicant receives a lengthy report based on hundreds or thousands of hours of evaluation of the company or organization. Prior to this, there is a process of guided self-evaluation which will help any company establish a baseline for its quality efforts and set a course towards excellence in quality management.

THE AMERICAN SOCIETY FOR QUALITY (ASQ)

The American Society for Quality (ASQ) is a leading organization in the field of quality management. Its web site, *www.asq.org* opens with an excellent quality management primer, and the organization, as a whole, provides a tremendous amount of support for any company or organization seeking to improve its quality management using any methodology.

THE PROJECT MANAGEMENT INSTITUTE

The Project Management Institute (PMI), at *www.pmi.org* is a global professional association for the advancement of project management. Project management and quality management are closely allied fields. Both focus on meeting customer requirements. Project success requires good quality management. At the same time, any change to an organization—such as a quality improvement program—is best organized as a project or a set of projects. The PMI is also advancing standards at a more strategic level, including the Organizational Project Management Maturity Model (OPM3), which integrates the CMM concept of maturity with the capability to do good project management; program management, which focuses on large activities consisting of many projects and other related work; and portfolio management, which evaluates programs as an investment of limited resources. The PMI—both at the global level and also through its local chapters—is an excellent resource that supports companies in developing a results-oriented structure for delivering either internal change or new products and services to customers.

INDUSTRY-SPECIFIC STANDARDS

Every industry has its own technical quality standards that should be addressed by any quality management program. Some—such as the manufacture of food and drugs—are regulated by law. Others have standards that, while optional, are beneficial because they exceed legal requirements in a way that helps improve business. Two standards organizations are of particular interest, ANSI and IEEE. The American National Standards Institute (ANSI) at *www.ansi.org* is a not-for-profit organization that guides the development of American national standards. These standards are open, meaning that they are developed using a collaborative, balanced, consensus-based process. ANSI coordinates with many other agencies that develop standards. When the standards are mature and stable, they are adopted as ANSI standards. For instance, the PMI's Guide to the Project Management Body of Knowledge (PMBOK) 3rd edition is the current ANSI standard for project management practices.

The Institute of Electrical and Electronic Engineers (IEEE) is an excellent organization that does more than its name implies. For example, its standard for a software engineering requirements specification, IEEE 930-1993, is an excellent guide not only for software engineers, but for the design of any specification, including all quality specifications and most business documents. You can learn more about the IEEE at *www.ieee.org*.

Conclusion:
Does Certification Improve Quality?

Just as a focus on quality should support—rather than be a sideline to—meeting core business goals of customer satisfaction and profit, so certification, or registration, or applying for an award, or developing a connection with a professional association should support improved quality and business success, and not be pursued for its own sake. That is, any of these activities should be aligned with corporate goals and evaluated for its value to the bottom line.

That said, many companies will find value in using one or more of these resource. It is simply less expensive to buy—or even pick up for free—the wheel of continuous quality improvement than it is to reinvent the wheel yourself. I do not mean that a quality management system can be purchased off the shelf and installed. That simply won't work. A deep willingness to change one's own philosophy and character, and to lead others to do the same, is essential to quality management success. But guidance and support for that process can be found in standards, awards, and professional associations.

Q-Ball Quiz for Chapter 12

1. Which of these is *not* true about ISO 9000?
   - (a) ISO 9000 is the European quality standard.
   - (b) ISO 9000 guidelines match the initial focus of a TQM initiative.
   - (c) ISO 9000 focuses on quality audits and quality assurance more than on quality control.
   - (d) ISO 9000 standards and their variants are used in various industries worldwide.

2. Which of these is *not* true of an ISO 9000 internal auditor?
   - (a) The auditor reviews procedures to ensure that documents (SOPs) are being followed and that records confirm this.
   - (b) The auditor must be trained in ISO 9000 auditing.
   - (c) Although the auditor may do other work in the company, he must confirm his independence in relation to each area he audits.
   - (d) To maintain independence, the auditor must work only as a full-time auditor.

3. ISO 9000 requires
   (a) inspection to ensure customer satisfaction.
   (b) use of statistical quality control.
   (c) either inspection or statistical quality control, whichever is appropriate to your business.
   (d) an annual internal audit program reviewing all ISO 9000 requirements.

4. Which of the following is *not* a criterion of the Malcolm Baldridge Award?
   (a) Leadership
   (b) Strategic planning
   (c) Quality management
   (d) Customer and market focus

5. Which of these standards, prizes, awards, or associations would *not* be applicable to any North American company seeking to improve quality?
   (a) The Deming prize
   (b) ISO 9000
   (c) The Baldridge Award
   (d) IEEE

Six Sigma

[W]ithin a few months after starting, enough knowledge had been obtained to much more than pay for the work of experimenting.
—Frederick Winslow Taylor, *Scientific Management,* 1911.

Six Sigma is the hottest approach to quality management in manufacturing and other businesses in North America. It is not a standard like ISO 9000; it is a methodology—as is TQM. As a result, individual companies can choose it, but it won't be imposed by corporate customers. ISO 9000 grows throughout an industry as major companies or governments require it of vendors. Six Sigma sprouts up here and there, like mushrooms in a forest overnight. Sometimes, Six Sigma finds good soil, fits well, and helps a company improve quality and achieve business goals; other times, it slips into a quality quagmire. If we understand the history and core concepts of Six Sigma engineering, we can implement Six Sigma successfully in our businesses.

A History of Six Sigma

Six Sigma was a direct outgrowth of North American TQM in the 1990s. TQM succeeded to a large degree. It saved major American manufacturing companies from even worse disaster. And the companies that adopted it soonest gained the most. Ford moved ahead first. Chrysler waited until Lee Iacocca came in and pulled off his famous turnaround. GM experimented by letting the independent Saturn division lead the way as a pilot project, then adapted Saturn's proven practices for its other divisions. Xerox lost industry leadership, but became a solid company and survived. American paper companies followed suit. Companies in a number of other industries—from UPS in shipping to Frito Lay in snack foods—improved through TQM.

When TQM did not deliver as promised, it was usually for one of two reasons. Internal failure was the result if the company slipped on one of Deming's 14 points—especially in failing to adopt the new philosophy with constancy of purpose or with training in leadership and quality throughout the organization. In other cases, TQM was achieved as a comprehensive practice but it delivered less than expected because competition didn't stand still. Every American car company moved to TQM, so the improvements led to survival, but not to greater market share. And Japanese companies—particularly in the automotive and electronics industries—didn't stand still.

As a result, by the 1990s, becoming a TQM company, maintaining 3 sigma quality levels, and extending TQM up the chain to suppliers and ahead to customer service was enough to stay in business, but not enough to attain or maintain leadership. Two companies that saw this early on were Motorola and GE. They picked up on Joseph M. Juran's realization that TQM with only elimination of special causes would lead to initial improvements, yet would eventually stagnate into doing things the same way over and over.

Here's how that works. If a process is not initially under control, then instituting measurement, control charts, and PDCA to eliminate special causes found on the control charts will solve a lot of problems of production process quality, significantly reducing defects and increasing productivity. If this is extended to all operations in the company, and non-essential operations are eliminated, additional value is realized. At that point, the company is at three sigma. It will make gains by extending TQM up the supply chain to vendors and forward to customers. But then there is nowhere else to go to make improvments.

These companies were looking for a direction that would give them a breakthrough as Juran recommended and Deming foresaw would be essential. The notion of a breakthrough as an essential ingredient for business strategy appears in a great deal of management literature. Related topics are the BHAG (big hairy audacious goal) suggested by Collins and Porras in *Built to Last: Successful Habits of Visionary Companies* (1994) and widespread techniques for stimulating outside-the-box thinking, creativity, and innovation. In the 1990s, Motorola and GE were already moving to 4 sigma and 5 sigma levels of quality control in some business areas, and 6 sigma was a big step in the next direction.

What made Six Sigma such a big step? Imagine that you've just produced one billion gazingus pins, or computer chips, or potato chips, or anything else for sale. Take a look at Table 13-1 to see how many defects you have in them at each level of sigma, and what percent of remaining defects you have to eliminate to move from the prior sigma level to this one.

Because the normal curve is not linear, each level of sigma is harder to achieve than the one before. Also keep in mind that the fewer errors that remain, the harder it is to hunt them down. That is how Six Sigma can be both a next logical step and a breakthrough goal.

Obviously, Six Sigma in the statistical sense is only relevant if you have millions or billions of events or products to measure. In this context, it is limited to the manufacturing of millions of identical products or billions of identical components, and is very well suited to the electronics industry. The methods, however, can be applied at other scales. They are particularly valuable when creating systems that put human life at risk, such as airplanes and cars. Six Sigma can also be applied to transactions, such as large banks processing millions of checks per day.

The core of Six Sigma engineering was developed at Motorola, GE, and a few other companies. Six Sigma became a movement when Jack Welch, GE's legendary CEO, announced an initiative to bring all divisions of the company to the

Table 13-1. Sigma levels and defects remaining.

Sigma Level	Defects per Billion	Percent of Defects to Remove to Move from Prior Level
6	2	99.7%
5	573	99.1%
4	63342	97.8%
3	2,700,000	94.1%
2	45,500,000	85.7%
1	317,000,000	n/a

Figures are rounded to three significant digits

Six Sigma level of quality. At that point, Six Sigma moved from an engineering approach to a quality management system in its own right, and also, unfortunately, became a confusing buzzword as well.

The confusion is created by the fact that Six Sigma can be defined as a goal (measured in number of defects per million opportunities), or as a management or executive initiative that includes commitments to Six Sigma or other breakthrough goals, or as a set of processes to achieve those goals. At Motorola and GE, Six Sigma succeeded because it was all of these. But efforts to market Six Sigma to other companies have often held just one or another piece of the picture. Let's look at components and processes of Six Sigma in depth, and then clarify some of this confusion when we evaluate Six Sigma in comparison to TQM and other methods.

Variations on Six Sigma

There is no single definition of Six Sigma, nor a central standards or certification organization. Companies do not get certified for having Six Sigma programs or achieving Six Sigma quality levels. Individuals can get certified, as Six Sigma black belts (able to manage Six Sigma projects) or green belts (able to support Six Sigma projects). Perhaps the choice of a martial arts metaphor in the qualification process is appropriate. In the martial arts and in Six Sigma, there are many different schools and the quality of training programs and real ability of certified people varies a great deal. Having a black belt doesn't mean that you will win a Tae Kwon Do competition or that you will improve quality for your company. It just gives you a fighting chance.

Just as martial arts training teaches many techniques, Six Sigma training includes many analytic and statistical processes. Paul Keller's *Six Sigma Demystified* presents over 50 tools, each of which can be used in one or more of the stages of the Six Sigma methodology. Each time a Six Sigma project manager or engineer runs into a problem, he has a tool in his kit.

The lack of standardization is a concern for executives and managers who want to know that the programs they choose and the people they hire will deliver desired results. The only effective solution is for Six Sigma—like TQM and ISO 9000—to be led by executives who clearly understand and are committed to Six Sigma principles and who assure education for everyone involved.

Flexibility is a plus side to the lack of standardization in Six Sigma. Companies can develop their own flavor, adjusting six sigma to meet the needs of their industry and their unique situation and culture. Indeed, this has happened, resulting in a lot of variations in six sigma methodology. This has allowed Six Sigma to move beyond its manufacturing roots into service industries, and also to be applied as an approach to process improvement where measuring billions of identical items is not the main issue. This flexibility comes with a risk: There are only two people who can determine if an expert made the right choice in choosing a process. One is another expert, and that is expensive. The other is the customer. And if the customer finds the error, that's even more expensive.

Six Sigma Simplified

The combination of diverse, company-specific approaches and advanced statistics makes Six Sigma very complicated. We can simplify it by looking at four elements: principles, components, processes, and measurement.

PRINCIPLES OF SIX SIGMA

These principles can be extracted from texts on Six Sigma:

- *Senior executive support.* Note the distinction between support, required for Six Sigma, and leadership, required for TQM.
- *Top-down training.* As in any quality management improvement effort, the need for training must be evaluated, and enough resources provided.
- *Include the voice of the customer.* This is a reminder to make sure that improvements really benefit the customer so that we avoid meaningless change or change that goes in the wrong direction because it is based on what we think the customer wants.
- *Create an infrastructure to support success.* The organization will need a structure that integrates Six Sigma methods into processes and allows discoveries through Six Sigma analysis to lead to process change.
- *Develop short-term projects with specific goals.* This is perhaps Six Sigma's most significant innovation. While some TQM companies naturally discovered the value of setting goals and deadlines, the project-oriented approach became a specific component of Six Sigma. Projects can have both minimum goals and *stretch targets* which motivate the team to think outside the box.

- *Focus on process improvement.* Just because Six Sigma uses projects to achieve results does not mean that the end of the project is the end of the improvement. Project definition is often based on an evaluation of the process identifying defined, measurable elements that are *critical to quality (CTQ).* Project results are usually internal changes—process improvements that should be maintained by ongoing Six Sigma quality control.
- *Clear and consistent methodology.* Although there are many variations of Six Sigma, a very consistent approach must be developed by each business.
- *Decisions based on fact and data.* Six Sigma reasserts the importance of an empirical basis for decisions, just like every quality management movement since Taylor in 1911. Greg Brue, in *Six Sigma for Managers* (2002) emphasizes this with these directives on page 45: "Ask questions. Challenge answers. Put assumptions to the test. Confront conventions."
- *Focus on people and processes.* Six Sigma realizes that our team is a corporate asset we need to invest in, and that team can only benefit the company if it can improve processes by defining CTQ elements that make changes to quality, time, or cost in business processes, products, services, and the bottom line.

COMPONENTS OF SIX SIGMA

Here are components critical to Six Sigma success:

- *Executive leadership and support.* Although some advocates say Six Sigma only requires executive support, in practice, the success of Six Sigma requires executive leadership and ongoing commitment.
- *Standard methodology.* The overall methodology must be defined, and everyone must be trained.
- *Projects to improve processes.*
- *Controls to keep improvements in place.*

When we create challenges by setting breakthrough goals and support and train people with good critical thinking skills in making lasting changes, we create a culture where a Six Sigma initiatives can really boost our company.

SIX SIGMA METHODOLOGY—DMAIC

The core of Six Sigma methodology is DMAIC, a minor modification of PDCA which, when spelled out, looks like this:

- *Define.* We need to turn customer requirements or executive directives for improvements into clearly defined goals, and use a project charter to define the project's purpose, business needs objectives, stakeholders, team members, and sponsor.
- *Measure.* In this stage, processes and CTQ factors are identified, and the current situation—our starting point for the project—is defined. Logical elements of factors are defined, establishing the factors critical to the success of significant business measures. Processes are defined in such a way as to allow effective statistical measurement, and the initial measurements are made.
- *Analyze.* We then analyze work processes to identify improvements.

  There are three main areas to analyze:
  ○ *Value stream analysis* is borrowed from lean manufacturing, which we will discuss in Chapter 16: *Gemba Kaizen for JIT.* The focus is on eliminating waste, especially wasted time, by eliminating unnecessary steps and improving poorly designed steps.
  ○ *Analyze sources of variation* using statistical process control techniques.
  ○ *Determine process drivers,* that is, figure out which inputs or variables change the CTQ elements of a process in key ways. This is done through advanced statistical techniques such as regression analysis, and by designing and performing experiments.

- *Improve.* First, we define a new process that we think will be better than the current process. Then we evaluate its benefits and look closely at how it might fail so that we don't make things worse. Then, with executive approval, we implement the new process, improving the way work is done. Lastly, we verify that the new process is working. If it fails, we either fix it or roll back to the process we started with.
- *Control.* To maintain control of the new process: we document the process; we eliminate sources of human error; we provide updated training; and, we bring the new process under statistical process control by updating control procedures and eliminating special causes of variation. When the new process is stable, we document the actual business value achieved. If goals were not met, we begin the DMAIC cycle over again as corrective action. We create a lessons learned document to assist with continuous improvement.

There are a number of variations on DMAIC, some with a different acronym. Some are better for specific industries. If you decide to adopt Six Sigma, be sure to have your own black belts evaluate alternative methods and choose the best one for your company.

DMAIC is an excellent process. While not actually adding anything truly new to PDCA, it does clarify some points so that, when properly implemented, it avoids these pitfalls.

- When PDCA is applied only to special causes, it plateaus after initial benefits. DMAIC, by allowing executives to set breakthrough goals and by including lean manufacturing concepts, keeps the continuous improvement cycle improving.
- Identification of critical success factors becomes critical once TQM initiatives are through the first or second phase. When we start TQM, looking at things from the perspective of a new philosophy makes a number of valuable changes obvious. After achieving quality goals, we turn our attention to any obvious weak spot—poor quality from suppliers, customer delight, slow cycle time, or high production costs—and solve those problems. At that point, TQM will plateau unless we find out which subtle changes will give us the biggest bang for the buck. Six Sigma takes us to the next step by providing tools to identify critical success factors.
- A focus that explicitly begins each project with a business goal that could be quality, time, or cost, but is always measurable links business improvement to technical improvement. The business goals are converted to technical goals in the Measure stage.
- The explicit inclusion of validation of both technical and business results ensures we don't make things worse.
- The explicit inclusion of developing and maintaining control of the new process prevents backsliding.

Six Sigma Measurement

Here we come to two critical points which are often misunderstood, and which create a lot of confusion. If Six Sigma had a different name, these would be problems only for engineers and statisticians. But, because Six Sigma is named after a statistical goal, lack of clarity about these two concepts ends up being critical to executive and management understanding and acceptance of Six Sigma as well. It turns out that Six Sigma is not as clearly and precisely defined as we would like it to be. To understand what slipped, we need to look at two issues: the precise meaning of defects per million opportunities (DPMO); and why the figure quoted as Six Sigma quality is actually equal to a much lower target—4.5 sigma.

DEFECTS PER MILLION OPPORTUNITIES (DPMO)

Six Sigma needed a new, more precise unit of measure. Quality levels four sigma and higher are above 99%. So we need to start measuring in millions instead of hundreds. But millions of what? That is the critical question.

To understand the answer, we have to look at the relationship between errors and defects. In the beginning of the book, I said we could treat them as synonymous until we got to advanced statistical issues. Well, here we are. Let's think of an error as something wrong with our process, and a defect as something wrong with our product. How are errors and defects—problems with product and process—related? The relationship is not a simple one.

- If our process requirements are not aligned with our product requirements, then we could have two problems. We could have a process that meets all our standards producing defective products. Or we could have an error in process that doesn't make the product defective. The first of these is solved through basic application of QA. The second is solved by techniques from lean manufacturing.
- Even when all process errors are significant to product results, the relationship between errors and defects is not one to one. Instead, it is systemic. A single process error—a defective piece of machinery or an oven at the wrong temperature—can turn out a whole batch of bad products. Or a flawed process can result in a defect that—perhaps due to redundancy designed into the product—is never detected by the customer.

These problems can be resolved, and are resolved, by quality engineers. They are resolved when we provide a precise definition to the term Defects per Million Opportunities (DPMO). The key question is: What is an opportunity? We have to look at this question both in terms of process errors and defects in products.

Let's look at DPMO in process first. In a process, any attribute of any step has an acceptable range of variance. Therefore, any attribute of any step is an opportunity for error. This includes each significant attribute of each input, each process step, each tool, each technique, each resource, and each method for validating output. Since the manufacture of any one item has hundreds of steps, each with a large number of attributes, there are many opportunities for error on the way to creating each product, component, or service interaction with a customer. An opportunity for an error in process is sometimes called an event, so you will also hear discussion of errors per million events.

We can also look at DPMO in terms of our product. Each attribute of the product has an acceptable range of variance. Anything outside that range is a defect. So an opportunity for a defect is our quantity of identical products times the number of attributes in each product. In reporting this number, we have to identify the number of defective products and also the number of defects, because one product could have several defects.

The fact that "opportunity" or "event" has an ambiguous definition is an unavoidable problem with no quick fix. It is inherent in the nature of systems—and our process and product are a system—that the relationship between errors and defects will be complex. On an engineering level, we solve the problem by defining the term opportunity in a way that keeps our eye on the goal.

The problem becomes large in the management reporting of our status. A single airplane is made up of over a million parts, each with many attributes. What is the measure of defects measured in DPMO? Is one successful flight one opportunity for a crash? Or is it three—one on takeoff, one for problems in flight, and one for landing? Or is one opportunity one critical part that might be defective? Or one error in manufacturing or maintenance processes? Unfortunately, Six Sigma initiatives can sometimes become rife with distorted and meaningless statistics when the answers are presented to show impressive success instead of to do good quality management.

WHY SIX SIGMA = 4.5 SIGMA

The most common description of Six Sigma is "a goal of fewer than 7 defects per million events (DPMO), and the quality management process that will get us there." In addition to the challenge defining DPMO opportunities or events, there is another problem with this definition: 6.7 DPMO is only 4.5 sigma quality. Six Sigma quality is over 1000 times higher, about two defects per *billion* (British—thousand million) events. How did such a big slip-up in the definition of Six Sigma happen?

It all started at Motorola. An engineer named Bill Smith—considered the father of Six Sigma—found that, at Motorola, internal quality levels of statistical Six Sigma—two defects per billion opportunities—resulted in external failure rates (customer quality levels) around 4.5 sigma—seven thousand defects per billion opportunities, or seven defects per million opportunities. He called this the *1.5 sigma shift,* and attributed it to a difference between short-term internal QC measurements and long-term product results.

If his investigation into the cause of the 1.5 sigma shift has ever been published outside of Motorola or verified, I have been unable to find it. The 1.5 sigma shift may be rooted in a fundamental statistical principle. If so, it would be nice to understand that principle and how it applies in different settings. It may be a rule of thumb that applied only at Motorola, or only in electronics manufacturing, or only in manufacturing and not in service industries. And we simply don't know.

The 1.5 sigma shift is an assumption that has become a part of the popular and widely accepted definition of Six Sigma. I actually believe this to be a serious problem. It means that there are a lot of Six Sigma business consultants, advocates, and engineers who are violating their own fundamental principles. Questioning everything—eliminating rules of thumb and replacing them with definable, measurable results—is at the core of Taylor's Scientific Management, of TQM, and of Six Sigma. If the definition of Six Sigma contains an unexamined rule of thumb, one has reason to question the entire system. How many other rules of thumb are hanging out where they don't apply? How many methods are being misapplied? How often is hard data being replaced by convenient guesses that are easier to obtain? There is no way to know—and that hurts the credibility and potential value of Six Sigma.

Evaluating Six Sigma

Six Sigma has been called "TQM on steroids," and the metaphor may be apt in both positive and negative ways. Six Sigma tools certainly add a lot of punch to TQM. They allow for audacious goals and provide continuous improvement. In a TQM environment—that is, where the executive leadership has held the TQM philosophy for a long time and Deming's 14 points are solidly in place—Six Sigma adds great value. There is no question that at Motorola, GE, and other companies that thoroughly embraced TQM and succeeded, Six Sigma has added lots of value.

Why then doesn't Six Sigma call itself an extension of TQM? For a very simple reason. Once Six Sigma succeeded at GE, it needed some other place to go. Manufacturing companies in trouble were a logical place to start. Unfortunately, by the mid-1990s, U.S. manufacturing companies in trouble all had one thing in common—TQM had failed for them. The companies that refused TQM altogether were long gone. The companies that succeeded with TQM were either using Six Sigma or had adopted their own methods and didn't need help. So the primary target market was executives who had to choose between thinking that TQM was a failure and seeing that they had failed to implement TQM. The easier sale was to agree with the executives—TQM was a failure—and now we have a new solution, a solution that is working at GE and Motorola: Six Sigma.

There is a problem with that approach. If TQM failed, there must have been a reason. If that reason isn't identified, it might well not be addressed. If it is not addressed, then it will cause the Six Sigma initiative to fail just as surely as it caused the earlier TQM initiative to fail.

Sometimes, companies for whom TQM failed succeed with Six Sigma. There are a number of reasons. Being in bad trouble a second time may be enough of a shock to cause a real shift in philosophy the second time around. Or maybe the executives who failed with TQM retired or were booted out. Or maybe a really good Six Sigma consultant succeeded where the TQM consultant failed. For any of these reasons, a company that faltered in a TQM or other quality initiative may succeed at Six Sigma.

Unfortunately, there have been many failures as well. Why? There are two reasons:

- Six Sigma focuses on executive support, not executive leadership. But executive leadership is really needed. Without it, the crucial change of philosophy may not occur.
- Six Sigma acknowledges human factors, but doesn't go far enough. Most particularly, Deming's eighth point, *drive out fear* from the workplace is not a formal part of Six Sigma, but where fear is present, process improvement will be blocked by a culture that allows fear.

As a result, there is a good analogy between Six Sigma and steroids, although it is a bit complicated. In a company that does not adopt the TQM philosophy and drive out fear, Six Sigma is like steroids for an athlete—it promises great improvement, but makes him less intelligent and creates health problems. Where fear is present, Six Sigma will fail for the same reason that quality assurance failed back in the 1970s, and the same reason Deming left the U.S. for Japan—engineers will be unable to convince executives to do good quality management and change processes.

On the other hand, when executives live the TQM philosophy, they are ready to take care of the company and heal its problems. In this environment, Six Sigma is like steroids used by doctors. The proper dose of a powerful drug, carefully controlled, in the right place and time can heal illness and boost a person—or a company—back to excellent health and better-than-ever performance.

Conclusion:
The Six Sigma Breakthrough Strategy

To get the most from a Six Sigma program, we need to eliminate the hype and see what Six Sigma has to offer—total quality management with a well-structured

process and extra-powerful analytical and statistical tools. If corporate executives see this, then Six Sigma can be applied in a larger philosophical framework—most likely TQM or *gemba kaizen* for lean (or just-in-time) manufacturing—and really add value. If Six Sigma is adopted by executives hoping that expert consulting can replace sound leadership, then their companies will suffer from their lack of willingness to examine their own limitations and their own crucial role in corporate success.

Six Sigma fits well with American business culture. Six Sigma black belts are the Texas Rangers of quality management—faster, better, tougher than anyone else, but working for the good guys. The focus is on innovation and breakthrough—a signature quality of American business. Six Sigma turns individual heroes into team leaders with special training. It is very interesting to contrast this approach with the Japanese approach of *gemba kaizen* for lean manufacturing (see Chapter 16), where the focus is on steady progress in small steps by everyone in the company, plus occasional breakthrough innovation.

Q-PRO

The Six Sigma Approach at Your Company
Given that Six Sigma truly means quality at a level of fewer than two errors per billion opportunities, it seems like it wouldn't matter for most companies. Unless you are producing millions of products, you are unlikely to make billions of measurements.

That is why Six Sigma is an unfortunate name for a potentially excellent approach to quality management. DMAIC, as a method, offers a lot of value to any company through the integration of business and customer requirements into defined technical goals; the emphasis on defined projects that create improved processes; the ability to look at each problem from so many angles and with so many tools; and the insistence on validation and maintenance of the new process. These tools could be valuable for your company even if you have only a dozen, a hundred, or a few thousand customers. If so, forget the name and use the tools!

Q-Ball Quiz for Chapter 13

1. The true statistical value of Six Sigma quality is
   (a) fewer than 7 defects per million events.
   (b) about 2 defects per million events.
   (c) fewer than 7 defects per billion events.
   (d) about 2 defects per billion events.

2. DMAIC stands for
   (a) Develop, Meter, Assess, Initialize, Conform.
   (b) Draconian Managers Anger Intelligent Computer-geeks.
   (c) Define, Measure, Analyze, Improve, Control.
   (d) Dubious Methods Aggravate Intelligent Critics.

3. Two key companies involved in the development of Six Sigma were
   (a) Motorola and GM.
   (b) Motorola and GE.
   (c) GE and GM.
   (d) Ford and GM.

4. Which of these is *not* a key benefit of executives setting stretch or break-through goals?
   (a) Customers get excited about the company.
   (b) Engineers are pushed to think outside the box.
   (c) Stretch goals, when achieved, put the company further ahead than incremental improvements.
   (d) Without stretch goals, quality management effort tends to level off.

5. Which of these is the most accurate statement about the relationship between a project and a process in Six Sigma?
   (a) A project becomes a process when it is maintained over time.
   (b) A project improves a process, and then the process must be maintained at a new level.
   (c) Projects are implemented only when there is a problem with a process.
   (d) Six Sigma focuses on the processes inside a project.

The Cost of Quality

As yet there has been no public agitation for "greater national efficiency," no meetings have been called to consider how this is to be brought about. And still there are signs that the need for greater efficiency is widely felt.
—Frederick Winslow Taylor, *Scientific Management,* 1911.

This chapter is about quality and the business bottom line. We've already discussed—in fairly precise terms—how improving the quality of our processes both improves product quality—leading to customer delight, lower sales costs due to repeat sales, and a better shot at increased market share—and also reduces waste in the production process leading to reduction in cost. More sales at lower costs per sale is a good one-two punch for knocking out executives who resist efforts to improve quality. Now, we will take a closer look at the idea of the cost of quality and how to calculate the benefits of quality.

This chapter also explores the benefits of quality guru Philip B. Crosby, possibly second only to Deming in his contribution to North American quality management in the second half of the twentieth century. Philip Crosby's ideas—the zero defect movement, the cost of quality, do it right the first time, quality is free, and hassle-free management—sound like buzzwords, but are actually succinct expressions of sound quality management principles. Crosby made many independent discoveries that paralleled TQM, and both had similar critiques of American management.

Life Cycles and Total Cost Models

There are two fundamental ideas in the cost of quality model, plus a little bit of simple math.

- *The total cost of quality is the cost of the effort to eliminate errors and defects, plus the cost of defects that remain.* That is, when we spend money to prevent or remove a defect, we save money at the other end, when the customer gets a working product.
- *Prevention costs less than design review; design review less than inspection or QC; inspection or QC less than letting the defect reach the customer.* This is the practical consequence of the 1:10:100 rule for the ratio of cost of defect elimination through planning, developing, and delivery that we discussed in Chapter 2: *The Development of Quality Management.*

This means that we can determine how much money we can spend on quality management before we begin a new program or project. See the simplified case study in the sidebar.

Q-PRO

Calculating the Cost of Quality

Let's say that your company sells 100,000 gazingus pins at $1.50 wholesale ($2.99 retail, less 50 cents internal retail cost, for $2.49 retail after cost of sales) per month, and that 25% of your sales are direct retail sales to customers. Let's also say that you have already achieved 3 sigma quality levels in your end product—good work! When a defective product reaches the customer, though, you refund their money and also send a replacement, at a total cost (including work time) of $12.00. You estimate that the market for gazingus pins will hold steady for the next ten months. A consultant comes to you and says that he can bring your company up to 4 sigma level and that, when he's done, operating costs will be identical to what they are now. How much should you be willing to pay for this project? Take a look at Table 14-1, which shows how to calculate the answer.

Table 14-1. Case Study: Calculating the Cost of Quality.

	Total Quantity	Quantity Defective	Item Cost	Replacement Cost	Total Cost of Defects
Current		270			
wholesale	75,000	203	$305	$2,436	$2,741
retail	25,000	67	$134	$804	$938
Total					$3,679
After the project		6			
wholesale	75,000	5	$8	$60	$68
retail	25,000	1	$2	$12	$14
Total					$82
Cost savings per month					**$3,597**
Cost savings over ten months					**$35,970**

To interpret Table 14-1, keep these points in mind:

- Three sigma quality is about 270 defective gazingus pins per 100,000.
- Four sigma quality is about 6 defective gazingus pins per 100,000.
- You sell 75,000 gazingus pins wholesale, and 25,000 retail, and we will assume that defects are evenly distributed across wholesale and retail. Fractions are rounded to whole numbers.
- Item cost for wholesale is $1.50, and for retail is $2.00, because the cost of sale is lost as well.
- Total replacement cost, including refund, is always $12.00.

Based on Table 14-1, you can see that you will save about $36,000 over the course of ten months. In deciding the value of the project and how much you are willing to pay for it, you would also consider:

- The risk that the project might not achieve desired results.
- The time value of money—how much is $3600 ten months from now worth to you today? Of course, this is a bigger issue if we are projecting years into the future.
- Soft dollar value, such as: the benefit of having even fewer dissatisfied customers—greater good will—which can be turned into reduced sales costs and possible growth; or the possibility that the new skills your team learns will lead to opportunities to reduce cost of production.
- Other scenarios, such as possible changes to the size of your annual sales.

In a cost of quality calculation, we compare the costs of our quality project or program against the benefits that result from having higher quality—realized as the benefits of fewer defects downstream.

Philip Crosby

Philip Crosby was a quality manager working for an aerospace company developing missiles for the U.S. government in 1961 when he had the Aha! moment that began the Zero Defects movement. His team had just delivered six missiles that were inspected by the customer. Nine to twelve defects were found and fixed in each, then all six missiles test-fired successfully. Everyone was happy with the tests but unhappy with the defects, and Crosby's boss demanded more. He wanted a perfect product delivered to the customer.

Crosby protested, but his boss insisted. Crosby brought the problem back to his team. One of them protested, but he insisted. And they sat down and got to work. That was when Crosby realized that defects aren't allowed if we make a commitment to being error-free. We get important things right. We allow defects when we think it doesn't matter—or when we've convinced ourselves that an error-free product is impossible.

ZERO DEFECTS

Setting zero defects as a goal has real value. If we set any level above zero as a goal, we're saying that we're allowing errors—no, really, we're planning to let error in. The goal of zero defects is a breakthrough idea to change that mentality. The term "zero defects" irritates statisticians who think that we can only approach an absolute, but never get there. But Crosby's environment—delivering a sophisticated product in batches of six at a time—was not an environment where statistics applied. Each part could be made right the first time. Each part could be assembled correctly. And zero defects could be achieved by good design and good process keeping out defects from the beginning. That's the idea of Get It Right the First Time.

To understand the technical issues and the approach to research and redesign for elimination of defects, see the discussion of seams and the weight of parts in Chapter 8: *Quality Engineering*.

HASSLE-FREE MANAGEMENT

In 1979, Crosby formed his own quality management training and consulting company. He developed his original ideas further, focusing on the fact that most quality problems happen because senior executives put up roadblocks of all sorts, getting in the way of effective process change. That is, corporate culture is full of hassles and elements that de-motivate people—that is, barriers that make people want to give up. Crosby proposed that the elimination of hassle and the elimination of defects were one and the same.

Here are Crosby's 14 steps to hassle-free management and zero-defect quality:

1. *Management commitment.* Crosby means executive commitment, from the top down, stated in a quality policy.
2. *The quality improvement team.* The quality improvement team has access to and full support from executive management. Their focus is improvement and they have the ability to clear away roadblocks.
3. *Measurement* of inputs, processes, and outputs.
4. *The cost of quality* is defined formally and objectively, and linked into the corporate accounting system. The cost of quality is the key measure of quality management activities.
5. *Awareness* of quality issues must be made simple and clear so everyone can understand it. This includes use of training and speeches, but also clear and impressive visual demonstrations.
6. *Corrective action* is not just fixing things after the fact. "The real purpose of corrective action is to identify and eliminate problems forever." (*Quality Without Tears*, page 112, 1984)
7. *Zero defects planning* focuses on the full launch of the new approach: a Zero Defects Day where executives announce their commitment.
8. *Employee education* gives people the tools of process change and quality management, and also leads them to expect that all hassle can be eliminated by process change.
9. *Zero Defects Day* is the executive commitment to continuing to move towards and maintain zero defects.
10. *Goal setting.* Define major goals and put target dates on them.
11. *Error-cause removal.* Have people report errors and their causes. Then the quality team guides people in removing the causes.
12. *Recognition* is given to those who are selected by all the employees as leaders in quality improvement.
13. *Quality councils* bring quality experts from across the company together to support one another and define next steps.

14. *Do it all over again.* After two years, a quality team may feel that everything has been done. Those folks are ready to drop off the team and focus on their regular job full time. One stays on to guide an entirely new team, with entirely new directions for improvement.

The Cost of Quality in Any Company—Including Yours

Crosby has an interesting view of manufacturing and service companies. He can't tell the difference. A factory mechanic putting an engine in a car is performing a service for the customer by making a working car, and a bank manager is performing a service for a customer by assembling the components of a new mortgage or loan. In discussing manufacturing and service industries, Crosby says, "The real problem comes about because the perception exists that office work, or functions like marketing and employee relations, can't be accomplished to procedures and specifications. Therefore, they have the privilege to be sloppy if they want to be." (*Quality Without Tears*, page 123, 1984)

Q-PRO

Quality in Personal Life

Crosby believed that quality management goes far beyond business. To illustrate his point about how we can achieve zero defects when we think it is important enough, he tells how he transformed his habits to save his own life. After a heart attack, he lost 20 pounds on his doctor's orders. Six months later he gained it all back, and realized that he had a quality management problem. He measured his intake and exercise, defined a program, and then stuck with it with the discipline of preventing weight gain by not putting unnecessary food into his body. His choices required culture change— new habits—followed by management style—keeping his commitment to his new way of life.

To perform cost of quality analysis, we need to use financial tools such as return on investment (ROI), and business planning tools, such as a thorough understanding of our project and product life cycles. We may need to do genuine research to gather information that has been hidden or disregarded. Crosby stresses communication with customers, internally with the company, and with vendors. Gathering this data, we can define the cost of quality in the cost of lost sales, the hassles of fixing broken things late or for the customer, and so forth. Once we determine the cost of quality, we can define the value of improved quality management.

We also use communication to achieve zero defects. This communication is focused on defining inputs, and then asking the manager of the preceding process to deliver outputs that are zero-defect inputs for the next process. If zero defects is the expectation of everyone in what they receive from everyone else, errors are prevented. The question we ask our customer is: What would be zero defects for you? The questions we ask ourselves are: What can I do to deliver zero defects? What do I need from my supplier (internal or external) to do just that?

Conclusion: Counting the Cost of Quality

The most direct value of the Cost of Quality approach is that it gives us a way to explain to fiscally oriented executives how our work in quality management will benefit the company's bottom line. But there is also another benefit. Cost of Quality analysis can be used to straighten out a quality management program gone awry. Sometimes, quality management training and activities become a sideline that just wastes time and adds hassle without producing real change. In these cases, an evaluation of the program can lead to repair that will bring quality management back in line with corporate goals and back in control of corporate processes.

In relation to this, I want to clarify one point that Crosby makes repeatedly. I agree with the point but, as stated, it would often be misunderstood in the present day. Crosby says that "quality is a process, not a program." He means that quality management must control and change business processes, not be a separate program for motivational rah-rah or theoretical concepts with no application. I agree entirely. In current terms—using the Project Management Institute's definition of programs—it would be clearer to say that quality management is a process, not just a program. Specifically, a program—a series of projects—is used to establish quality management and then to change processes—but the process of ongoing maintenance of zero defects is as important as the program and projects that produce the change. When we realize that zero defects is a reality we can create in a hassle-free environment, we improve the bottom line, our processes, and our working life all at the same time.

Q-Ball Quiz for Chapter 14

1. The total cost of quality equals
   (a) the salaries of the quality management department and the time spent by employees on quality issues.
   (b) the cost of the effort to eliminate quality, plus the cost of the defects that remain.
   (c) the benefit we gain by implementing a better quality program.
   (d) the cost of the effort to eliminate quality, minus the cost of the defects that remain.

2. Which of the following is *not* included in the cost of a defect?
   (a) The cost to the customer of dealing with a defective product.
   (b) The cost of resolving the customer's problem.
   (c) The cost of scrapping the item.
   (d) The cost of preventing the defect.

3. Which of these is *not* a phrase used by Crosby?
   (a) Do It Right the First Time
   (b) Zero Defects
   (c) Eliminate Numerical Quotas
   (d) Hassle-free

4. In discussing American slogans to be eliminated, Deming said, "How could a man make it right the first time when the incoming material is off-gauge, off-color, or if his machine is not in good order?" Crosby is not recorded as answering this particular challenge. Imagine that he did. Which of these would he be most likely to say?
   (a) "You are right. We can only get it right the first time when we can report problems with machinery, processes, and inputs and get them solved. Upper management must give managers and workers the means to change processes to solve quality problems."

(b) "You miss the point. Do it right the first time assumes that everything is in working order. It's just about paying attention."
(c) "With diligent effort, these problems can be addressed."
(d) "A quality management program would solve these problems."

5. Crosby's view of manufacturing and service companies is
    (a) quality management is easier in manufacturing companies.
    (b) quality management is easier in service companies.
    (c) quality management works in both types of companies, but it has to be done differently in each.
    (d) there is no real difference between the two.

15

The Capability for Quality: CMM and CMMI

In the future it will be appreciated that our leaders must be trained right as well as born right, and that no great man can (with the old system of personal management) hope to compete with a number of ordinary men who have been properly organized so as efficiently to cooperate.
—Frederick Winslow Taylor, *Scientific Management*, 1911.

The Capability Maturity Model is an industry-specific quality model for software development. Software development is a huge industry and, in general, is rife with both project management and quality management problems. The popular term for software defect—a bug—has become a household word and an accepted fact. New Microsoft operating systems are released well after scheduled delivery dates, sometimes with tens of thousands of bugs. And that's not to criticize Microsoft—they work very hard up in Redmond, WA, and they are doing the best that they know how to do. Similar problems will be found on a smaller scale in many software development companies, and also in the many in-house computer programming shops that build software for internal use in large and medium-sized companies.

There has got to be a better way. And there is. A small number of companies are out there producing high-quality software and, when needed, zero-defect software. Their methods show a much lower cost of quality than most of what is going on in the industry. And their methods are a direct descendant of TQM. The situation in computer programming—another term for software development—today is very similar to the situation in assembly-line manufacture in Taylor's day 100 years ago. There is a better method, but few people are using it. The reasons organizations and software development teams fail to adopt best practices are centered in cultural resistance and poor communication, because these methods are all available for free.

This chapter looks at the methods of zero-defect software development, focusing on the Capability Maturity Model. In addition, we will see how the concept of capability and maturity have far-reaching implications for the future of quality management.

The History of CMMI

The first recorded structured effort to develop zero-defect software began with Michael Fagan, a certified quality control engineer at IBM. He learned Total Quality Management from Deming and Joseph M. Juran, and asked if the same methods that were working for the manufacturing industry might apply to the development of software. Juran, a TQM pioneer and colleague of Deming's, told him that the answer was absolutely yes, and that all he had to do was to apply TQM principles to his own field.

ZERO-DEFECT SOFTWARE

Michael Fagan's team got to work in 1972, and the zero-defect software initiative was born. Philip Crosby also consulted with IBM for many years, but I'm not sure if he and Fagan ever crossed paths. Software development is an unusual industry. Building software is much more like building a single prototype of an airplane than it is like manufacturing multiple similar or identical copies of a product in production or mass production. Unlike physical products, computer programs, once complete, can be replicated perfectly with almost no defects.

Yet each computer program is made up of thousands or millions of unique lines of computer code, and a single logical error, incorrectly assigned variable, or simple typo is a defect that might cause the entire program to fail. In that way, writing a bug-free computer program is a lot like building the prototype of a sophisticated airplane. A large computer program is a single system with a huge number of unique parts, each of them critical to the whole. But getting a computer program ready for use is like getting just one airplane ready for use. The problems of reproducing many identical defect-free copies—a major source of headaches in manufacturing—do not apply. Statistical quality control is useless. That is what makes Juran's assurances so interesting. Many people identify TQM with statistical quality control. But even when control charts do not apply, TQM methods do work.

Fagan's team created a method called *Software Inspection,* sometimes also known as Fagan's Inspection or the IBM inspection method. If you want to learn about it more fully, *Software Inspection* by Gilb and Graham (1993) is an excellent book. It was the first textbook on Fagan's methods, which he published outside IBM in 1976.

Software Development Methodology: An overview

Before we take a look at Software Inspection and other quality management best practices in the software development field, let's take a look at some of the problems in the way software is sometimes written today.

- A programmer gets what he thinks is a good idea, or gets a request from a customer.
- He starts to play around, writing computer code and making something.
- He puts it together into a package, designs some tests, runs the tests, fixes what he can, and lists the rest of the problems in a readme.txt file.
- He delivers a beta or the final product.

If this is what is going on at your organization—and a lot of programmers and web developers work this way—then you need project management before quality management. Just as Taylor and others organized work gangs into assembly lines, so software development needs a structured process. Programming as described above is like sending a construction crew to an empty lot with only the instructions, "Build a house." If a house gets built, it won't be the one you wanted, and it will cost a lot more than you planned.

Project management solves that problem. One of its most useful tools is the idea of a life cycle—a series of stages with gates at the end. Work is done in stages, and gates are when we review or, more formally, inspect the work before moving into the next stage. Setting up stages and gates catches and fixes defects early, preventing defects from moving ahead and taking advantage of the 1:10:100 rule to reduce the cost of quality. Software Inspection, in addition to applying quality management principles, also works within a structured life cycle. A structured life cycle provides stages and gates, and also allows decisions that must be made early—such as which of many alternative architectures to use for a new software application—to be locked down before the design work that depends on architecture begins. Here is an example of an effective software development life cycle (SDLC) summarized from my book *Project Management Demystified:*

- *Create and evaluate concept.* Somewhere, someone—a customer, an IT expert, or a senior executive—gets an idea that a new or improved computer program would be good for something. The idea could be seen as a new opportunity or as the solution to a problem. It could be novel or copied from somewhere else. Wherever the idea comes from, it needs to be evaluated. Is it a sound idea for the business with a good return on investment and acceptable risk?
- *Analysis to define requirements.* As discussed in Chapters 3 and 4, we begin by defining customer, stakeholder, and technical requirements. In many software development efforts, this should actually be the bulk of the work. In software development—perhaps even more than in other industries—most errors come in at this stage if they are not prevented.
- *Define architecture.* Information technology—just like construction—needs architecture. Architecture is the ability to design something while seeing it several ways at once, taking into account constraints imposed from each perspective. In software architecture, users want all requested features on clear, efficient, easy-to-use interfaces. Computer capacity, operating systems, and other programs impose limitations and require interfaces. Selecting the best components and tools for building the program, and defining the overall structure of the program, is the work of computer architecture. In selecting an architecture, current knowledge and tools is a factor due to the cost of education and retooling, but should not be a controlling factor.
- *Design tests.* Tests should be developed directly from the requirements specification, either independently of, or before, the writing of the computer program.
- *Specify the computer program.* This is the technical stage.
- *Write and revise the computer program, its documentation, and instructions.* This step should include low-level review or inspection and low-level testing. Work should be organized using the project management method known

as *Work Breakdown Structuring* and modules should be defined and developed in an orderly fashion.

- *Put the program together, with inspection and testing.* Extensive testing to ensure that the program works to all requirements and is acceptable to the customer is done in a stepwise fashion as each module is finished, and then two more series of tests are run on the complete product, one before the customer receives it, and one after.

- *For published software, release a beta test version.* Software published into an open environment where there are many different models and versions of equipment, operating systems, and other software cannot be proven to be compatible with everything that is out there until it is tested by users. A good beta-test program begins with defect-free software that needs to be tested for fit with the user's environment, and engineers resolve compatibility issues before preparing the final release.

- *Final release and customer support.* The final release includes extensive product, user, and technical documentation. Training or support for training is almost always essential to product acceptance and success. Customer support includes correction of defects not found previously, resolution of new compatibility issues that arise, and documenting user requests for new functionality in preparation for new development efforts.

Even when a well-structured SDLC is in place and project management methods are used, there are still legitimate barriers to the entry of quality management. Software engineering is a complicated field where the tools rapidly change. This creates three problems:

- Engineers and programmers spend more time learning new tools, and have less time to learn and apply quality management.
- The divide in understanding between computer programmers on one side, and customers or executives on the other, is wide, making the translation of customer requirements into technical requirements, and the validation of that translation, very difficult.
- Any quality management or project management method will need to be frequently adapted to new tools. Prototyping tools—which allow a programmer to show users a mockup of a program very early—change implementation of life cycles. Automatic software testing tools change implementation of quality management.

This last point is key. What software developers need to realize—and many do not—is that, while new tools do change the implementation of project management or quality management in their field, they don't change the fundamental principles of the field, or replace the value of quality management with an automated solution. As always, successful quality management is an application of universal principles, customized to a unique situation.

Software Inspection is a solution to these problems, and can open the door to development of high-quality software much more rapidly and at lower cost. Software Inspection was developed for large teams on projects many months or even several years long. Other methods have adapted the same core quality principles to smaller work groups. Let's look at Software Inspection, and then at other methods.

The Software Inspection Methodology

Software Inspection grew in a way parallel to other quality management methods. Developers first focused on their own work—computer programming and debugging—and then moved the process up earlier in the supply chain to design, and then tied everything together. The result is that the full method—called Software Inspection—contains much more than inspection (close independent review) of software. Let's look at Software Inspection as a whole quality management practice designed to eliminate defects at the earliest possible point, from beginning to end.

Software Inspection is not active in the concept stage. The end of the concept stage delivers a project charter that assigns a project manager, defines the objective or goal, and sets a due date. From that point forward, the software development project manager using Software Inspection will follow a highly structured process.

A structured series of defined processes (SOPs) will already be in place, defining each task with its input documents, process instructions, and desired outputs. In software development, an output is often a new, improved, or more detailed version of the input. As a result, version control—tracking what steps have been taken on each document—and document management—keeping track of the documents themselves so we don't end up with multiple versions—are essential.

Table 15-1 introduces the four core tools of quality management for software development, and explains the function of each one. Although only one is called "software inspection," the entire Software Inspection approach to producing high-quality software includes all of these steps.

Table 15-1 is a valuable tool in explaining to software developers why they need inspection even if they are already using walk-throughs, reviews, and testing. Table 15-2 makes it even clearer by comparing the process of developing software to the process of developing a prototype airplane.

Table 15-1. Major components of software inspection.

Component	Quality Function	Stages	Benefits
Walk-throughs	QA—makes sure everyone understands enough to contribute best ideas	Analysis, architecture, design, and whenever new stakeholders or team members are brought in	Better specification and architecture reduce total cost and cost of quality
Reviews	QA—experts contribute to design, applying cross-functional skills	Some analysis, especially architecture and design	Designs quality in
Inspections	Inspection of component outputs at every stage of the process (inspection or QC)	All	Identifies and eliminates defects early in the process, prevents components with defects from moving forward before they are fixed
Tests	Inspection or QC	Development and later stages	Testing of components in operation, and in environments as close to actual use as possible

Table 15-2. Quality activities—comparing software development to airplane prototyping.

Component	Airplane Prototype	Software Application
Walk-throughs	Make sure everyone understands what the plane is supposed to do	Make sure everyone understands what the software is supposed to do
Reviews	Get everyone expertise to make sure that the plane will fly	Get everyone's expertise to make sure the software is well designed and will work to specification
Inspections	Look closely at every part to catch problems before the test flight	Look closely at every component to catch problems before the test run
Tests	Test the parts of the plane and then the whole plane in operation	Test components of the software and then the whole program in operation

No one would doubt that each of these processes was essential in making a new airplane, and no test pilot would fly a new airplane unless he was darned sure that the plane had gone through all the earlier steps with flying colors! But, in software development, it is all too common that essential quality processes are skipped, or they are done incompletely, or even that they identify defects, but the defects are not isolated and repaired before the next step is started. Or, sometimes, the defect is repaired, but the module is not fully retested. This is bad news, because studies of software development show that one new bug is added for every five or six that are removed. Let's take a closer look at the new process that Fagan added to software development: Software Inspection.

Software Inspection Process

Picture the following situation. You have just finished a piece of work—let's say working out the detailed definition of a module or code object based on the specification. This is still design work, before programming. You know that all the inputs you used were already inspected. You followed a step-by-step process (SOP) in doing your work. You checked it as carefully as you could. Then you gave it to everyone else on the technical team, but not the managers. Everyone on the team—perhaps 50 people—has read your document. A facilitator—with help from recorders and others who keep track of everything—has called an inspection meeting.

In the meeting, each line of code is read aloud. Anyone is invited to comment on each line. All comments are recorded and discussed. As this is done, suggestions are made to improve the design document that you created. But that's not all: Every single document and process is subject for review. For example, if your document has a defect, the work is traced back. If you followed the procedure correctly, then the inputs are rechecked. If the inputs are valid and you followed the procedure, but created a defect, then the procedure was wrong. Not only will you correct the problem in your module, but someone will be assigned to correct the procedure so that it will work from now on. At the end of the meeting, you feel you've been put through a ringer. But you haven't—only your work has. And now you are given your work with a list of things you can do to make it come out with zero defects. You are in charge of making your own work as good as it can be. The team has helped by finding the defects long before testing or even development of the code. And you are asked if you need any help or training to get it right.

In this way, every output and every process are continually inspected. The inspection meeting is the core of the process. To make the process work, however, it is essential that:

- Every meeting be fully documented.
- All nonconformities be isolated and fixed before the component is allowed to become an input to the next step.
- All recommendations for further review or to change SOPs or other documents be tracked, assigned, and completed.
- Appropriate additional meetings be held as needed.
- Although quality control charts are not applicable, statistical metrics of process—hours of effort on review, for instance—and results—defects found and corrected—are essential to collect and to use in the evaluation. That evaluation rolls into the PDCA cycle for continuous improvement of the software, the methodology, and the Inspection process itself. As we approach zero defects, the evaluation can be supported by the seeding technique described in Chapter 10: *Statistics for Quality.*

As you can see, Software Inspection is a huge effort. It's value was publicly demonstrated when IBM used it to produce the zero-defect software that launched the Space Shuttle, managed its orbital operations, and brought it home for a landing. In fact, when Nobel Prize winning physicist Richard P. Feynman evaluated every critical component of the Space Shuttle after the Challenger disaster in 1996, he concluded that the software was the only component that had been designed and developed to ensure zero-defect quality. It is ironic that the industry with the worst reputation for quality developed the most reliable piece of the shuttle, while aerospace engineers faltered in other ways.

Software Inspection became publicly available because NASA's mandate is to make all its methods available for general industrial and economic use. The result of Software Inspection at NASA was the CMM methodology. Before we go there, let's briefly identify some other valuable software development methodologies.

Other Effective Methodologies and Tools

The only big drawback of Software Inspection is that it is big. It works for large teams and requires an investment in training. Even at NASA, the question was asked: How can we apply the same principles and achieve zero defects, but do it at a lower price?

Key principles of Software Inspection include:

- Following a structured methodology. Structured methods exist for requirements elicitation, requirements definition, applications design, and program coding.
- Having every procedure fully documented.

- Early and independent design of checking through inspection and testing.
- Use of requirements tracing.
- Independent checking of each procedure and component.
- An environment that thoroughly reviews the work in a way that supports the workers.

Three other methods have been shown to be rigorous enough to provide quality management that moves a software product towards zero defects:

- *ISO 9000.* ISO 9000 can be adapted to software development.
- *Creating a Software Engineering Culture,* a flexible approach for small teams that comes out of Kodak in Karl E. Wiegers's book of that title in 1996. Weigers also wrote the Microsoft Press book, *Software Requirements.* Both books offer extremely practical approaches to organizing a small software development team for high-quality work. He addresses very well the issue of whether it is worthwhile to get certification, or simply to develop best practices.
- *Egoless programming* is an expression of the idea that there will always be errors, and it is better to have the team find them than to have the customer find them. I prefer the notion that it takes a strong ego to do quality work, but we should leave our ego at the door when we walk into an inspection meeting—whether we are reviewing someone else's work, or having our own work reviewed. We need a cooperative focus on doing good work to eliminate and prevent defects.
- *Buddy programming* is an innovative approach that actually has two people working simultaneously at one computer screen to write code. As one writes, the other debugs and documents. The two switch off writing, letting whoever has the better idea move ahead. If the observer doesn't understand what the writer is doing or thinks the writer is off track, they talk about it.

There is a universe of good tools available. A proactive quality manager, project manager, or lead programmer can identify appropriate best practices and lead a team on a process of quality improvement in software development. The reduction of hassle and increase in pride of workmanship make it worthwhile for the team. The reduced defect rates make it worthwhile for the customer, and for the company who gets more reliable systems and has to fix a lot less bugs later on. If you want to move ahead on improving software quality, Wiegers's books are an excellent place to start.

NASA, SEI, AND CMM IN THE UNITED STATES

An IBM group developed the Space Shuttle software for NASA using Software Inspection. A think tank, the Software Engineering Institute (SEI), was contracted to identify best practices in software development from NASA and the U.S. military and make them available to the public. The SEI, operating as a think tank, stepped back and took a very large look at the issue of what it takes for an organization to be able to produce defect-free software. They looked beyond methodologies to the larger issue of capability, asking: What makes it possible for an organization to be able to use a good methodology and get good results? In doing so, they developed the Capability Maturity Model (CMM), a model that suggests that organizations have a defined, measurable level of maturity in process, project, and quality management that allows the organization to deliver results. If an organization wants to improve substantially, it must develop a program to develop its maturity, then implement that program and follow through to maintain a mature status of capability. This defines an idea that appears throughout this book very clearly—the idea of a quality improvement program as a temporary endeavor that changes processes and culture, and then the new culture becomes self-maintaining at the new level—and puts that idea as a central requirement for maturity and capability. Also, it makes quality improvement programs replicable. With CMM, it is possible to define where you are, define where you want to go, and create a program to get there all following other companies who have done it before you. The Capability Maturity Model was developed over the course of five years, beta tested at a few companies for two, and was in use as a standard for certification from 1992 until 2001, when it was replaced by the newer Capability Maturity Model Integration (CMMI).

Although CMM was developed in the software industry, the five levels of capability are, in fact, appropriate to all organizations—corporate, governmental, and not-for-profit—in any realm of activity:

1. *Initial.* Practices are not documented, or documented practices are not used. Work practices are based on individual preference or interpersonal training that is not standardized. Sometimes, this kind of on-the-job learning is referred to as folkloric transmission, as opposed to structured education. The organization is relying on unique skills of particular individuals, which puts development at risk if this person leaves the project or the job. Project organization depends on the skill level of the project manager. Estimation tends to be poor.

2. *Repeatable.* Within departments, project management and control are implemented. Methods are defined. Projects are tracked with metrics. Use of historical data improves estimation. However, different methods are in place in different parts of the organization.

3. *Defined.* Standard, documented processes for software development are used throughout the organization. Project management methods are standardized. An organization at this level could probably receive ISO 9000 certification with relatively little effort. A group, such as a quality team or project management office, has been created to facilitate the organization's process activities. An organization must reach this level before advanced quality control techniques can be applied.

4. *Managed.* Quantitative quality goals are defined and measured. For example, software inspection could be applied to ensure use and improvement of standardized methods. The group is able to define nonconformities—in process, product, or control—and use management initiatives to resolve them. Software development is of reliably high quality.

5. *Optimizing.* Processes for defect prevention, technology change management, and process change management are added. This leads to sustained continuous improvement.

These terms seem quite abstract in their formal definition. Here's what it looks like in plain English:

1. *Initial.* Each person learns and does the job any way he or she can. The organization relies on heroes.

2. *Repeatable.* Departments have gotten their act together, established procedures, and are following them and keeping them up to date. But methods are different in different places for no good reason.

3. *Defined.* The whole organization has standard methods for standard activities, and departments have unique procedures where they are needed. This is all centrally coordinated.

4. *Managed.* We're doing good quality management to eliminate defects. We're hit and miss on actually making things better.

5. *Optimizing.* We are using, sustaining, and improving best practices. We have continuous improvement rolling, we're using breakthrough approaches, or both.

One key factor should be highlighted: education. People always need training to be able to follow SOPs, training in their quality management role, and overall guidance in participating in the organization's continuous improvement efforts. However, in software development, this is especially true for several reasons. First of all, new and updated tools are constantly coming into play, both for development and for quality management. We need to integrate these tools with current systems, revise SOPs, and train people in the new procedures. Secondly, software development is a project-oriented field, and new project teams are assembling all the time. We must use training to bring the team together and assure everyone can follow the procedures. Thirdly, software development is a highly mobile profession. As we bring new people into the organization, we need to be able to bring them up to speed.

The five CMM levels apply to any industry but, of course, specific best practices vary from one industry to another. The original CMM became known as SW-CMM (for software), and other versions were developed for other industries. This is one of the crucial ways that CMM is different from ISO 9000. In ISO 9000, an organization has to have defined practices, but does not need to adopt any specific practice. CMM requires specific best practices for your industry for certification at a particular level.

Achieving a given level in the original CMM model was challenging because the system was monolithic. To get certification at any CMMI level, you had to demonstrate that the organization was at that level for all activities—specification, design, and so forth. That structure was appropriate for large-scale mainframe development environments.

CMMI—ADDING FLEXIBILITY

With the development of networked computers of varying sizes, projects and teams have become smaller. This was another driver in the development of CMMI. The problem with CMM certification being monolithic has now been addressed by the development of a new standard, CMMI. The word integration refers to the fact that the system has been made modular—split into pieces—so that it is easier to integrate with your organization and with other standards or methods you may already be using.

The modularization of CMM into CMMI is a good thing, except for the resulting barrage of technical terms. Here are the modules, which are called disciplines in CMMI lingo.

- *Systems Engineering* is about total systems, which may or may not contain software. It addresses issues of customer requirements and constraints.

- *Software Engineering* is for software development—writing computer programs.
- *Integrated Product and Process Development* addresses communication and coordination among stakeholders, and is used in conjunction with one or more other disciplines.
- *Supplier Sourcing* addresses how to ensure that suppliers provide what is needed for your development projects.

Full definitions are available at *www.sei.cmu.edu/cmmi*, a well-organized web site that guides you through understanding CMMI and how to implement it.

An additional issue to address is the process by which your organization will develop its maturity. Two approaches—called representations—are offered:

- *Continuous* representation allows you to pick the components you will develop and the order you will deploy them. It is more flexible, and also simplifies integration with other standards, such as Electronic Industries Alliance (EIA) 731.
- *Staged* representation is more structured, and is a path already followed by other companies. It is also the easiest path from CMM to CMMI.

In choosing a representation, be sure to find out if your customers—or potential future customers—require one or the other. Other criteria are available on the CMMI website.

CMMI Certification

It is perfectly reasonable to learn CMMI and adopt those parts that work for you in whatever order you like without ever seeking certification. Any organization wanting to implement CMMI should either hire people who developed and used CMMI in other environments, or retain consulting services. CMMI, because it requires industry best practices, has a steeper learning curve than ISO 9000. One consultant, when asked about developing CMMI without guidance, compared it to a little league team wanting to win the championship without using a coach. The analogy is apt: If you want to go somewhere, it makes sense to hire a guide who has been there. Many of the CMM methods will seem extremely alien to professionals who—by the standards of their profession—have been doing things the right way for a very long time.

Certification may be required by customers, and is certainly a selling point with new customers. The certification process is called Standard CMMI

Appraisal Method for Process Improvement (SCAMPI). The approach is similar to the one outlined for ISO 9000. An initial, quick evaluation, called a SCAMPI C, let's you know what you need to do. When you are nearly ready—really, when you think you are ready—you have a SCAMPI B, which is a more in-depth external audit. Final certification is granted through a SCAMPI A external audit.

There are two issues of concern with CMMI certification. The first is that, unlike ISO 9000, there is no process for maintaining certification. You are not required to develop an internal audit department, and there are no scheduled follow-on audits. Also, the group that is audited receives certification, but the company name can appear on the certificate. This was a major problem under CMM, where some large software development consulting firms would have one division that achieved CMM level 4 or level 5, and then use that to advertise the quality of its services. But there was no guarantee for the customer that the group fulfilling their contract had ever attained that CMM level at all.

These situations are somewhat ameliorated, as the SEI now maintains a database that will identify which group within a company was certified to what level on what date. However, this is not a full solution. The customer will still be uncertain that, as the group or team changes, it has maintained its CMMI level. And CMMI does not require the level of records retention—auditable proof that SOPs are being followed—that ISO 9000 does. As a result, a company may be in compliance with CMMI, but be unable to demonstrate that to a client through external audit. Any company or division that wishes to prove that it is operating at a particular CMMI level has to go through an expensive certification process, rather than a relatively less expensive recertification audit.

CMMI Compatibility With Other Standards and Methods

CMMI is highly compatible with ISO 9000, and even more compatible with Six Sigma. ISO 9000 provides a basic quality management framework, CMMI provides a software-development specific framework, and Six Sigma can be used both to focus the project on goals that add corporate value and also to accelerate the project. Six Sigma is more compatible with continuous representation deployment of CMMI, because the organization can use Six Sigma methods to choose what modules to deploy where and in what order. Imposing a goal of certification may create conflict between the goal of certification and other organizational objectives.

A MATURITY MODEL FOR PROJECT MANAGEMENT—OPM3

The project management institute has released OPM3, the Organizational Project Management Maturity Model. Using the five-level maturity model from CMM, it addresses the question of how capable an organization in any industry is of defining, developing, and delivering projects on time. The primary tool of organizational project maturity is the project management office (PMO), also called a project office (PO). The PMO usually does not run projects. Rather, it develops and improves the organization's best practices library and supports other divisions in learning, adopting, and using the library to achieve project success. Project management and quality management are closely related and even interdependent, as we will see in Chapter 18: *Practical Quality for Projects and Programs.*

CMMI Around the World

U.S. military contractors and some other government agencies are requiring their primary contractors to be certified at CMMI level 3 or higher. That requirement has recently been extended to include subcontractors as well, so smaller organizations are having to gain CMMI certification. There is specific information about CMMI for small organizations at *www.sei.cmu.edu/cmmi.*

Another major area of growth for CMMI is the People's Republic of China. Most major companies have adopted CMMI there, and the training and certification business is very large.

Conclusion: Evaluate Your Own Maturity

Whether software development is your thing or not, the notion of maturity levels is a valuable self-assessment tool for any business. In working on CMM and CMMI, the SEI has established documented proof that repeatable process—the core of quality management and the application of science to business—is indeed an essential basic requirement to delivery of quality results and to the ability to maintain and improve quality without backsliding.

So, where does your company stand?

Q-Ball Quiz for Chapter 15

1. Which of these shows the five CMM levels in correct order?
   (a) Optimizing, Managed, Defined, Repeatable, Initial
   (b) Initial, Improving, Defined, Replicable, Optimizing
   (c) Initial, Repeatable, Defined, Managed, Optimizing
   (d) Initial, Repeatable, Defined, Optimizing, Improving

2. Which of these is *not* one of the four core tools of quality management for software development?
   (a) Software Inspection
   (b) Project Management
   (c) Walk-throughs
   (d) Tests

3. In an industry where work is organized by time-limited projects, which is most true?
   (a) Project management and quality management are the same thing.
   (b) Quality management has to come before structured project management.
   (c) You can implement both quality management and structured project management at the same time.
   (d) Structured project management must be in place before quality management initiatives are effective.

4. Which of these tools is an inexpensive way to have independent review quickly on a small team?
   (a) Buddy programming
   (b) Software Inspection
   (c) CMMI level 3
   (d) Walk-throughs

5. Which of the following is the final audit that, if you pass, gives you CMMI certification?
   (a) SCAMPI B
   (b) SCAMPI C
   (c) SCAMPI A
   (d) Shrimp SCAMPI

CHAPTER 16

Steady Improvement in Japan: Gemba Kaizen for Lean (JIT) Manufacturing

*All of this requires the kindly cooperation of the management,
and involves a much more elaborate organization and system
than the old-fashioned herding of men in large gangs.*
—Frederick Winslow Taylor, *Scientific Management*, 1911.

"Improvement brings many truly satisfying experiences in life—identifying problems, thinking and learning together, tackling and solving difficult tasks, and thus being elevated to new heights of achievement." So says Kaizen pioneer Dr. Masaaki Imai, founder of the Kaizen Institute. And *kaizen* is the everyday Japanese word for improvement.

In Japan, the efforts started in the late 1940s by Deming, JUSE, Toyota, and SONY Electronics, continue to this day. Progress has been steady and evolutionary, rather than revolutionary. Yet it has arguably been more successful than the efforts to improve quality management in North America and Europe. Imai seeks to explain the success of the Japanese approach in his two books: *Kaizen: The Key to Japan's Competitive Success* (1986) and *Gemba Kaizen: A Commonsense, Low-cost Approach to Management.* His explication of the Japanese approach and investigation into its uniqueness is grounded in over 50 years of work and leadership in quality management around the world. He lived through the development of TQM—usually called Total Quality Control (TQC) or Company-Wide Quality Control (CWQC) in Japan—and continues to pioneer these practices around the world.

Dr. Imai's thesis is that Japanese business culture focuses on Kaizen (literally Change-Good, generally translated as continuous improvement). He contrasts this with Western business culture, which focuses on innovation. We could say that Kaizen functions like gentle rain throughout the growing season, letting a forest grow steadily year after year, while innovation is like a huge thunderstorm in a desert, leading to a sudden burst of vibrant growth that slowly dies away when the rain stops. I choose such an intense metaphor to set the stage because it is hard to express the difference between the subtle and the dramatic. But that is the key difference: Are many subtle, small changes, and only occasional big ones a better investment of quality dollars than large investments in breakthrough technology?

Table 16-1, derived from *Kaizen* illustrates some of the key differences between Kaizen and innovation.

Dr. Imai does not recommend Kaizen in place of innovation. Rather, he believes that companies need both. With Kaizen only, there would be only slow improvement, and a company would fall behind when other companies innovated. With innovation only, a company makes sudden leaps, but then slips back because—as the auditors teach us—without attention, everything degenerates. The optimal approach—and the best companies in the world, Japanese or otherwise, do this—is Kaizen so that things are always run to standard and standards are incrementally improving, combined with innovation to leap ahead.

In 1979, Mankichi Tateno, then president of Japan steel works, formulated three TQC goals (*Kaizen,* p. 45):

1. To provide products and services that satisfy customer requirements and earn customer trust.
2. To steer the corporation toward higher profitability through such measures as improved work procedures, fewer defects, lower costs, lower debt service, and more advantageous order filling.

Table 16-1. Kaizen and innovation—key differences.

Issue	Kaizen	Innovation
Where most practiced	Japan	North America and Europe
Size of each change	Small	Large
Cost of each change	Small	Large
Number of changes	Many	Few
Source of change	Workers, guided by management	Technical experts, selected by management
Location of source of change	Gemba, the shop floor	Research laboratory
Effect	Long term, but not dramatic	Dramatic, but short term
Timeframe	Continuous and incremental	Intermittent and non-incremental
Effort	Collective	Individual
Source of knowledge	Conventional know-how and desire to improve	Technological breakthroughs, new inventions, new theories

3. To help employees fulfill their potential for achieving the corporate goal, with particular emphasis on such areas as policy deployment and voluntary activities.

It is clear that Tateno had the bottom line in mind. But note how much emphasis there is on people and process—on how changing trust levels and fulfilling potential and improving processes is the way to achieve corporate goals in this mission statement.

As you read on, it may be challenging to understand "What makes Japan different?" or "What makes Kaizen different?" One reason is that the approach to quality management in Japan is continuous, not revolutionary. In the West, we expect to find a difference marked by a name and a movement. In Japan, changes to quality management are marked not by new names and new movements, but by new tools plus an appreciation of the original sources of innovation. Imai credits Deming, JUSE, and the U.S. Air Force for the innovations that Japan has continued to develop since that time.

For those familiar with quality management, even the new tools will not seem new or different. Here is a list of Japanese improvements to quality management that will probably seem familiar:

- *PDCA at all levels.* Deming introduced the improvement cycle as connecting research, design, production, and sales. Japanese managers and workers iterated it to address individual quality and production problems.
- *Quality teams.* Japanese companies fostered teamwork within department and cross-functional teams.
- *"The next process is your customer,"* Ishikawa's phrase that became the basis of SIPOC.
- *The five S's of good housekeeping.* Sort, Straighten, Scrub, Systematize, and Standardize.
- *The seven new analytical tools,* all used in Six Sigma: relations diagram; affinity diagram; tree diagram; matrix diagram; matrix data-analysis diagram; process decision program chart (PDPC); and arrow chart.
- *Flow production,* where workstations are arranged in order of work flow.
- *Pull production,* where each workstation makes only enough output to supply input as needed for the next step.

All of these tools have been tried with some success in the West. In fact, Six Sigma uses just about all the tools that come from Japan, and adds a lot of heavy statistics, as well.

The crucial difference is one of flavor. In Six Sigma, everything is made to sound big, impressive, larger than life—in a word, innovative. Becoming a black belt requires lots of extra study and a head for statistics. In Japan, it is a source of great pride to work on the shop floor—and this is as true of managers and engineers as well as it is of workers and foremen. Most training is on-the-job training. Everyone is involved, and it is more important that everyone contribute to the improvement program than that one person make a real breakthrough in how everyone works. The lesson may be that, with or without innovation, we need the attention and intelligence of every worker to maintain quality and productivity and improve our standards gradually.

As you read this chapter, instead of trying to find the magic solution, look at what makes the difference between a master chef and a routine meal. Focus on the attention, detail, and thoroughness with which Japanese executives and management have brought quality management to permeate through their companies. That is the real difference.

We will look at three broad components of Japanese quality management: Kaizen, or continuous improvement; Gemba, or where the action is; and Just-in-Time manufacturing. For an indepth understanding of these tools to a level that would allow you to implement them, see Dr. Imai's books.

Kaizen: The Japanese Contribution to TQM

Dr. Imai suggests the following breakdown in the use of our time:

- *Executives.* About ⅓ innovation, ⅔ continuous improvement, or kaizen
- *Managers.* ⅛ innovation, ⅜ kaizen, ½ work
- *Supervisors.* ⅓ Kaizen, ⅔ work
- *Workers.* ⅛ Kaizen, ⅞ work

For this to happen, companies must set up training and structured systems for continuous improvement. Let's look at how this is done.

KAIZEN VISION FOR EXECUTIVES

Quality comes first, and profit second. Quality—having a longterm vision of serving society—is the reason for being in business. Staying in business requires QCS—leadership in managing quality, cost, and scheduling. Creating structures for the dissemination of the vision and strategy, and ensuring the collective of the entire company participates, is key.

Process (P) is emphasized more than results (R). Process is rewarded. For example, a group of cafeteria workers evaluated tea consumption in the cafeteria by serving it in large pots, and then seeing how much tea was left for several weeks. They used analytic techniques to track and measure tea consumption. Since people usually sat at the same tables, after a few weeks, they were able to deliver right-sized pots of tea to each table, cutting tea consumption in half. The cost savings for the company were minimal, but the company awarded the cafeteria quality team with a very high prize for their work.

P means people as well as process. The emphasis on guiding, supporting, and empowering people is key to kaizen.

KAIZEN MANAGEMENT

The focus of kaizen management is a complete appreciation that each error is an opportunity to learn and improve. In kaizen, the role of a standard is fundamentally different than the role of standards elsewhere. A standard documents the best way we know how to do something today—and therefore the worst way we should ever do it. In kaizen, we don't strive to meet standards. We meet them, and strive to make them a little bit better.

How much is a little bit? This will amaze you: In Japan, any suggestion that saves 0.6 seconds—one hundredth of a minute—is taken seriously, evaluated, and adopted if effective. How much benefit is there in that? A lot, if every worker is making hundreds or thousands of suggestions per year.

Suggestion Systems

Kaizen doesn't use a suggestion box; it uses a suggestion system. And it really needs one, too. Managing suggestions is a huge part of the kaizen effort. Workers are taught that suggestions that might have any of these benefits are valuable:

- Improvement in one's own work.
- Savings in energy, material, and other resources.
- Improvements in the working environment.
- Improvements in machines and processes.
- Improvements in jigs and tools.
- Improvements in office work.
- Improvements in product quality.
- Ideas for new products.
- Customer service and customer relations.
- Other.

Every suggestion is treated as a plan that kicks off PDCA activity. If a change is in one's own work, the manager is likely to approve it, and then the worker does it himself and the results are measured (checked). If the result is beneficial, the SOP is updated (action). Small work groups have the ability, skills, and tools to modify their own work spaces. Larger changes are led by management.

Each change also moves in another direction. It is tabulated as a statistic of the suggestion system, which is also subject to evaluation and improvement through PDCA. A suggestion management system evolves through three stages over a 5 to 10 year period. In the first stage, the key metric is the number of suggestions. The goal is to have every worker make suggestions, and the more, the better. Across all Japanese industry, the average is 19 suggestions per employee per year, 76% of which are put into action. But leaders in Kaizen management have many more. In 1985, Matsushita Corporation handled over 6 million suggestions—16,821 of them from one employee.

In stage 2, the quality of the suggestions is emphasized, and workers are trained in identifying changes that are likely to result in improvement. Only in stage 3 does management begin to focus on the bottom-line impact, encouraging workers to see the connection between each change and quality, cost, or scheduling (QCS) and evaluating the suggestion system for its results on the bottom line. And this can be five to ten years into the process. In the first several years, the focus is on the process of people being involved in the suggestion system, and measure of success is in followthrough on suggestions into action and overall tracking. There is a bottom-line benefit in results (R), but it is a side effect of focus on people and process (P).

EYE ON THE BALL

Suggestions Add Up
This is a summary of a case study of the suggestion system at Aisin-Warner from pages 114–119 of *Kaizen*. The company focused on supporting workers in filling out suggestion forms with small ideas that could be implemented at low cost. In 1982, they had 127 suggestions per worker, for a total of over 200,000 suggestions, 99% of which were implemented. The suggestions led to improvement in many areas: reduction of effort; quality; facility improvement; safety; materials savings; and more.

ALL THE ANGLES

Suggestions in America
Do this exercise to see if you can imagine the barriers to a kaizen suggestion management system in a North American company:

- *Propose it to executive leadership.* Would they believe they could get hundreds of suggestions per employee? Would they be willing to invest management time in evaluating tens of thousands of employee suggestions? Would they care about improvements to office orderliness or savings of 0.6 seconds of cycle time? How would they feel about investing even a small amount of money and time in a program where bottom-line benefits could not be measured for the first several years?
- *Imagine managing it.* Imagine you are in charge of 150 workers. Can you imagine handling 18,000 suggestions per year, and implementing over 15,000 of them?
- *Imagine how workers will feel* if they are encouraged to make suggestions, and rewarded for changing the workplace to make it better in any and every way.

KAIZEN WORKERS

The kaizen worker knows his job. When he is trained in the SOP, he knows he is not only encouraged to meet it, but to find ways to make it better. This creates a very different attitude than the attitude of the worker who thinks he has only been told how to do his job. It also creates a different attitude than what happens when a person is left to figure out the work for him or herself. The worker's attention is focused on the relationship between the work and the SOP. If he can't do the SOP, he tries harder or gets help. And help will come, because the quality control on cycle time will let his supervisor know if he's running behind. If he can do the SOP easily, then he has time to think about ways to make it better, and knows that there is a very good chance that his suggestions will be implemented, leading to pride and praise. The suggestion system allows for immediate approval of small suggestions, so he sees how, if he can figure out how to make his job easier, he can do it right away.

QUALITY TEAMS

Workers receive training and have more time to focus on improvements through participation in quality teams. There are two types of teams—departmental and cross-functional. The departmental teams seek to improve the operation of their own equipment and processes. The departmental team often develops in three stages. Kaizen begins with learning, and the team studies together. Then the team focuses on its own processes, and makes suggestions for improvements. The team meetings help with orientation of new team members also. Suggestions are on a larger scale than suggestions from individual workers, and often require months of analysis and effort. If the suggestion is approved, the team implements it or manages its implementation. This can lead to the third stage, where major improvements become a center of corporate volunteerism and team building. For example, one team came up with a significant modification to the operation of the furnace in their shop. When they proposed it, the welding shop team suggested that, instead of having the welding shop do it for them, they could learn welding themselves and modify their own equipment. They took to it the way American children take to the Little League. They re-engineered their own furnace on weekends and off-time—not for pay, but for pride in their work. And, of course, they had to learn a lot to do it. Kaizen begins with learning and ends with learning.

Quality teams—called QC circles in Japan—were originally voluntary, and management does not require participation. They grew out of the study groups formed by the first people who took Deming's classes around 1950. JUSE defines the QC circle as "a small group that voluntarily performs quality control activities within the shop where its members work, the small group carrying out its work continuously as part of a company-wide program of quality control, self-development, mutual development, and flow-control and improvement

within the workshop. By engaging in QC-circle activities, the circle members gain valuable experience in communicating with colleagues, working together to solve problems, and sharing their findings not only among themselves but with other circles at other companies." (Kaizen, pp. 161–162)

The quality team and the suggestion box can also become a career path. One way to move up is to become a roving suggestion-maker, moving through the entire company and making small suggestions for improvement everywhere.

Gemba Kaizen

If you try to reach a Japanese manager in his office, you will probably be out of luck. He is much more likely to be in *gemba*. Gemba is a Japanese word that literally translates as "real place" but has the connotation of being on the spot or where the action is. In Kaizen management for manufacturing, gemba is the shop floor. In service work, it is anywhere an employee might encounter a customer. In police work, it is the scene of the crime. In news reporting, it is live-action reporting.

A manager should be in gemba for the same reason that a detective should be at the scene of the crime. Problems are in gemba, and problems are opportunities for improvement. The core of gemba kaizen is to respond to every abnormality—problem or possible problem or even unusual improvement—with these five steps:

- *Go to Gemba.* Show up where the problem is so you can respond immediately.
- *Investigate Gembutsu.* Gembutsu are things in gemba—physical equipment and physical evidence, such as the defective product. Investigation means seeing and touching, not just hearing about.
- *Temporary countermeasure.* Kick the machine to get it running again.
- *Find and remove the root cause* using TQM tools.
- *Update the standard* to make sure that the problem—and others like it—do not recur anywhere in the company.

The manager's work in gemba kaizen focuses on process. Much of the work improves process in commonsense ways without worrying about the bottom line. More technical work is validated through improvement in process measures, such as reducing the mean human effort (measured in seconds of work per person) or machine time per unit produced. Finally, the manager links these activities to the results-oriented bottom-line measures familiar to Western business.

The manager's work in gemba kaizen also focuses on people. Developing teams, being present where there are problems to show caring, being present to approve immediate adoption of a good suggestion, leading movements that support safety and cleanliness, and supporting quality circles (learning and improvement teams) make up a good deal of the manager's day every day.

Just In Time (JIT)—Lean Manufacturing

In America, we often misunderstand the Just-In-Time manufacturing approach that was developed by Toyota. I have learned it as an alternative accounting method. Efforts to explain it as lean manufacturing come somewhat closer. But—as both Deming and Imai warn—lean manufacturing and just-in-time logistics management are impossible until each process within the company has been optimized and then standardized through kaizen. A new company might be able to adopt gemba kaizen for lean manufacturing all at once. An existing company should adopt gemba kaizen in preparation for a shift to lean manufacturing. Also, a company that has problems with absenteeism or employee morale can't make lean manufacturing work.

Toyota developed lean manufacturing in response to a crisis in its early years. It had produced cars and trucks for sale very efficiently using QCS, but then did not meet its sales projections. All the unsold vehicles meant a big loss that threatened the company's financial stability. Taiichi Ohno responded by committing to make only vehicles for which the company already had orders. Later, he said that this approach was essential to allow the company to make as many different models of vehicles as it does. And Toyota was one of the few automotive companies to remain profitable through the oil crises of the 1970s.

Just-in-time manufacturing is actually an application of a very basic TQM principle—fact-based decision making—at a very large level. How should senior executives decide how many of which items to make in the next year? Relying on sales projections is not fact-based, it is more like using a crystal ball. Sales projections are usually created by sales managers with little training in statistical forecasting. Worse, there are many incentives for bias—promising to sell a lot is how a sales manager sells himself to management. He can always get yelled at and blame the economy or the competition later when his projections are not met.

What if a manufacturing company said, "We won't build anything until we have an order for it. But we will streamline our operations to the point where we can get it to you so fast you will be happy." That is the basis of just-in-time manufacturing, also called lean manufacturing, or the Toyota method.

This decision drives a complete reorganization of the production process. That reorganization has three essential elements:

- *Optimal standardization with everything measured and under control.* For a new company, the whole operation is build on best practices, with quality control and process control for cycle time in place. For an existing company, these are achieved through gemba kaizen in preparation for the shift to lean manufacturing.
- *Workflow optimization.* The physical layout of the plant and the input and output specifications of each process are optimized for efficient workflow handling one piece at a time.
- *Pull production.* Each process is triggered by a request from the process that follows it in line. A process produces only as many items as are actually needed by the subsequent process, and they are delivered when needed.

Lean manufacturing has two major performance components: the elimination of waste, and the maximization of operating efficiency.

ELIMINATING WASTE

Muda, the Japanese word for waste, has a very wide meaning. Muda includes all items and activities that do not add value to the final customer product, including: overproduction, inventory, rejects, motion, processing, waiting, transport, and time.

As we've already discussed, the unit of improvement is either the $1/100^{th}$ of a minute (0.6 seconds) or the second. At a company like Toyota, a manager could tell you the exact number of seconds any part spent in the factory, and how many of those seconds were spent moving from one place to another, and how many were spent in a process that improved the product. He would hope that the number of seconds spent sitting still was very close to zero. While complete zero inventory levels are impossible due to unknowns, inventory is kept at an absolute minimum every step of the way. And inventory is not just counted as stocked items. If two items are sitting between one process and the next on the assembly line, that's inventory. And if it isn't there for a good reason, it shouldn't be there at all.

KEEP THINGS RUNNING BY SHUTTING THEM DOWN

This is a paradoxical approach that is based in the paradox that every problem is an opportunity. It also requires a crucial piece of kaizen philosophy—that everyone should admit his or her mistakes openly, immediately, and fully. We meet standards by recognizing when we have failed to meet them.

In gemba kaizen, machines are often designed to shut down immediately when a defective part is made. Workers always have the ability to shut down their own machines. And when a worker shuts down the machine, that's an abnormality. The manager shows up and does the five steps of gemba kaizen discussed above. When a machine is shut down, it is shut down for a reason. If gemba kaizen works, that only happens once. That reason will never be a problem again on that machine, or on any similar machine in the company.

The result is a paradox. The more often machines are shut down or operations are halted, the fewer repeating problems there are, and the more total productivity increases.

There is also a proactive side the gemba kaizen. One part of this is total productive maintenance (TPM), which involves management, workers, and quality teams in the effort to have a zero-breakdown rate. Another is support and management of workers leading to high attendance, attention, and productivity. A third is the effort to investigate things that seem wrong—dangerous, risky, or problematic—even when there is no clear problem.

If American auto makers adopted lean manufacturing, it would be a major culture change. Can you imagine an America with no year-end auto clearance sales to make room for the new models? In American business in general, the approach is to produce all we can, then figure out how to sell it. The lean manufacturing approach is different—it is based on Deming's original formulation of PDCA for the entire company: Find out what the customer wants, design it, produce it, and sell it. Let your sales and customer satisfaction surveys tell you how to change it, then act to make those changes in both quality and quantity.

Lean manufacturing seems revolutionary, and pull production is certainly a significant innovation. But it is an innovation that can only be achieved through evolution, not revolution.

Conclusion: Lasting Evolution

Quality management in Japan has succeeded by being evolutionary rather than revolutionary. The investment in small improvements by everyone is also an investment in every worker as an asset. Improvements are not driven by investment in innovation, or even by expensive bonuses or education for workers. Instead, inexpensive but personally significant prizes and gifts are the primary motivators, and practical education focuses on expansion of common sense into PDCA.

This approach applies both to improvements in quality and also improvements in quality management methods, as well. Japanese businesses are not looking for great new innovations, though it will be happy to evaluate them. Instead, they invest in micro-improvements at every level, and benchmark the progress of a continually improving quality management system—a kaizen system that is, itself, improved by kaizen.

Through kaizen, Japanese companies have become the first large group of businesses to empower managers and workers to apply Taylor's methods for improving efficiency. Quality teams are able to observe and measure processes, define improvements, and put them into place as Taylor taught his clients to do 100 years ago.

ALIGN YOUR Q

Awards, Standards, Customer Appreciation, or Business Success
There are four different ways to motivate a company to focus on quality improvement: We can seek to win an award like Japan's Deming Prize or the U.S. Malcolm Baldridge Award; we can seek certification to a standard; we can measure and seek to improve customer delight; or we can look at business results. The last two are probably always beneficial. The first two may help your company, or may not. And that is probably an issue of culture.

The Deming Prize has probably made more of a difference in Japan than the Baldridge Award has made in the U.S. And that may be a matter of culture. To win either prize, a company usually must apply for several years, getting better each year. Seeking external evidence of approval is very big part of the Japanese cultural identity. Continuing that effort for several years, knowing that one will probably eventually win, can focus a company for a long time. Americans may be more likely to be averse to failure, and to see not winning as a sign of failure. As much as we like competition, we hate to lose. So, in American business, we might do better seeking to meet a standard that is challenging, but not competitive, so we can be very likely to win soon—within the next fiscal year.

Q-PRO

Kaizen: How Successful?

There are two issues that raise questions about the Japanese management approach: Are we getting all the news? And, what about the Japanese economy?

All of the reports about Japanese quality management are positive. And that means they are certainly biased. The bias comes from four sources:

- Successes and benefits have been extraordinary.
- Other countries are not interested in Japan's problems and failures—we want to learn from their successes.
- Japan is, of course, interested in its failures. But the process of dealing with them—either becoming aware or hiding them, either taking action or not—is internal to Japan.
- It is probably the case that the widespread success allows failures to be fixed by processes that executive management or outside consultants see as routine. They simply need to apply expertise that has worked elsewhere. This is unlikely to be seen as newsworthy, especially if the repairs are incremental, unlike the newsworthy corporate rescues and takeovers of troubled companies in North America.

The result is that those of us outside Japan can neither easily evaluate the actual success of gemba kaizen for JIT manufacturing, nor learn from its mistakes.

The Japanese economy ran into its first long-term difficulties since World War II from 1995 to 2003, and it may not be out of the woods yet. What does this say about gemba kaizen for just-in-time manufacturing? Does the bad economic news mean that it isn't working? Is it part of the problem? Is it part of the solution?

Analysis suggests that Japanese operational management strategy does have significant effects on the economy, but that those are mixed in nature. The biggest factor is guaranteed "lifetime employment," which is a policy related to TQM. Internally, the plus side of guaranteed lifetime employment includes more value from long-term worker training and greater company loyalty. The down side is that a company cannot easily shrink, so production costs are likely to remain higher.

The reality is that only about one third of the Japanese workforce—mostly older men in large companies—have guaranteed lifetime employment. So the impact is not as large as it might seem. There is also a social issue: Is it better to pay idle workers or to have more unemployed people? One study suggests that, in a year when official Japanese unemployment statistics were at 3%, the idle workforce was the same size. Is Japanese society better off having those people be unemployed while businesses lower their prices to be more competitive on world markets, or retaining these people and facing a harder challenge in competitive pricing in the world market? It is difficult to know. The retention of employees adds stability; the inability to reduce workforce costs—the way Western companies can—could prolong the economic problems.

Another element of kaizen may be part of why Japanese industry has trouble snapping out of an economic downturn, though this is more speculative. Although Imai proposes kaizen—small improvements—plus some innovation for rapid change, many Japanese companies may have cultures that support only incremental change. In that case, the success of Six Sigma for major North American companies—which started at the same time as the Japanese economic difficulties—may mean that, at least for a time, Western innovation in quality management is proving better for worldwide business than Japanese kaizen.

What is clear is that the two have different functions. Continuous improvement—kaizen—is like steady athletic training to build strengths and eliminate weaknesses. It always helps the athlete. Innovation is like a new technique or technology—the Fosbury Flop of the 1968 Olympics or, more recently, innovations in uniforms that reduced air resistance around racing skater's uniforms. The question isn't, "Which is better?" The questions are:

- How can we create a culture that fosters both innovation and continuous improvement?
- How can we create a corporate culture that has room for best practices from both East and West?
- What is the right balance of investment in the two in quality management at our own company?

In considering these issues, it is important to realize that innovation can almost always be copied quickly, but that the asset of a quality workforce can only be cultivated and sustained slowly, though it can be lost quickly.

Q-Ball Quiz for Chapter 16

1. Which of these is a quality of innovation, and *not* a quality of kaizen?
   (a) An intermittent, non-incremental time frame.
   (b) Small changes.
   (c) Many changes.
   (d) Source of change is often the workers.

2. Which of these is *not* true when comparing kaizen to Six Sigma?
   (a) Kaizen improvements may not have a demonstrable effect on the bottom line, but Six Sigma projects must demonstrate that they will improve production or the bottom line to get approved.

(b) Six Sigma uses kaizen tools but also includes tools that require more education than kaizen would generally use.

(c) Kaizen quality circles operate in much the same way as Six Sigma project teams.

(d) Six Sigma and kaizen both seek to improve processes.

3. Which of these is *not* gemba as the term is generally used?
   (a) The shop floor
   (b) The scene of the crime
   (c) On location live-action news
   (d) A manager's office

4. Lean manufacturing is best described as
   (a) an alternative accounting technique.
   (b) the effort to reduce waste and inventory.
   (c) an effort to produce only the items ordered, supported by elimination of all forms of waste in a controlled manufacturing environment.
   (d) the industrial production of dietetic foods.

5. Shutting down a machine improves productivity in a gemba kaizen environment because
   (a) the machine is shut down when it is replaced by a newer, better machine.
   (b) shutting down a machine when there is any defect, then investigating the cause, leads to permanent preventative action so that that problem doesn't happen again.
   (c) the machine has to be shut down, because any worker who makes a mistake is fired, and there is no one to run the machine until he is replaced.
   (d) when the manager plays a prank by shutting down a machine, it increases worker morale.

PART FOUR

Practical Quality Management

Dan Millman, a colleague and expert in personal growth, likes to say that the difference between knowledge and wisdom is that knowledge is knowing what to do, and wisdom is doing it. In Part 4: *Practical Quality Management* I hope to help you take the understanding of quality management that you've learned by reading this far and put it to good, practical use. Whether you are a team member, a project or department manager, an executive, or a business owner, you can improve quality in the work you do.

Any effort to improve quality management itself—whether we call it a program, a project, a process, or an initiative—is an answer to a business problem. Therefore, that effort itself is subject to the PDCA cycle. We need to plan it, do it, check it, and act on the knowledge from checking to correcting, with feedback or breakthrough to the next level. In project management terms, when we schedule a problem for solution, that's launching a project. Part 4 offers guidelines for developing and conducting a program to improve quality management that leads to a permanent change in process plus the management support to continue to standardize and improve the new processes.

You may have noticed that each chapter of *Quality Management Demystified* begins with a quote from Taylor's 1911 treatise, *Scientific Management.* Those quotes are there to show that we are still figuring out how to apply ideas that were worked out in principle. When I first discovered how much Taylor had already figured out, I was irritated that, decades later, we are not using answers that are readily available. As I studied all of the good thinking and hard work that has been necessary to implement those ideas in the real world, I've moved from a place of criticism to one of admiration for all the good work we've done.

There's still more work to do. The goal of Chapter 17: *Challenges and Leadership* is to help you figure out what you can do to improve quality and quality management at your company. Chapter 17 takes a problem-and-solution approach. We define the reasons why—although we've known most of what to do since 1911—many of us have not been able to make it work yet. Then we'll see what business owners, executives, and organizational leaders in government, the not-for-profit sector, and education can do to overcome obstacles, set a course towards quality, and focus the team on improving quality. The approach in this chapter is equally appropriate to a first-time effort at quality management and to an effort to renew or improve the focus of a quality-oriented company using any methodology.

In Chapter 18, we turn our focus to *Practical Quality for Projects and Programs* where we learn how to apply best practices in quality management to temporary, time-limited endeavors with unique results. We focus on how to use quality management to improve the value of project results while reducing project cost. We also integrate quality management best practices with the Project Management Institute, *A Guide to the Project Management Body of Knowledge (PMBOK® Guide)—Third Edition,* which is the ANSI standard for project management.

We close *Quality Management Demystified* with a look towards our future in Chapter 19: *Global Quality in the 21st Century.*

17

Challenges and Leadership

This close, intimate, personal cooperation between the management and the men is of the essence of modern scientific or task management.
—Frederick Winslow Taylor, *Scientific Management*, 1911.

In spite of the facts that the essential principles of quality were developed in 1911 and rounded out by 1924, and that we've had solid quality theory available since the 1980s, truly high quality is more the exception than the rule in most of business. Why is it so hard for business—and society—to put these ideas into practice? We talk about barriers in corporate culture, but what does that mean? How come just a few businesses achieve quality in all aspects and maintain it? If we understand these questions and start to answer them, we can lead the way to improved quality management in our own businesses and organizations. In this chapter, we will:

- identify the barriers and challenges to success in quality management,
- identify critical success factors,
- outline a plan for a quality improvement program that will stick,
- discuss key factors for quality at the management and worker levels.

Solving a Problem That's Already Solved

Many people will tell you that a solution to the challenge of quality management already exists. I heard the founder of the *Kaizen Institute,* Dr. Imai, speak of *Gemba Kaizen* for JIT as a solution. Others say that Six Sigma is a solution. I both agree and disagree. Certainly, in some companies at some times, error has come consistently under management, quality has been a key focus of business leaders, and the drive towards quality has succeeded in delighting customers, delivering value, reducing costs, and increasing profits. And all of that is good.

At the same time, the challenge of achieving sustained comprehensive quality management is still with us and the problem is never fully solved. There are executives, and whole companies who don't care much about quality, or who give it lip service, or who think that they are managing and delivering quality when they are not. In fact, some of the companies that have the most trouble with quality management are those that have a system—such as TQM or Six Sigma—but have lost the crucial leadership and employee focus on excellence essential to maintaining a corporate culture that supports and sustains continuous quality improvements.

The fact is, in terms of ideas, the solution to quality management exists in many places. One can get there with TQM, with Six Sigma, with Gemba Kaizen for JIT, or by pursuing a standard or prize such as ISO 9000, CMMI, or the Baldridge Award. In a smaller company, we can achieve quality through caring about excellence and good team leadership, and then studying best practices and applying common sense, without too much worry about standards at all.

But ideas alone don't grow corn. To grow corn, we have to plant corn kernels, then pull up weeds. And weeds like to grow back. It's the same in business: A steady focus on quality will yield a rich harvest, but seeds of other things—such as pressure to increase profit, impress stockholders, cut costs, or keep things simple—will sprout and get in the way.

Barriers and Challenges

We often talk about resistance and corporate culture, but if we want to bring these issues under management, we need to define them more clearly. In *Kaizen,* Dr. Masaaki Imai defines corporate culture as "factors of industrial structure and psychology that determine the company's overall strength, productivity, and competitiveness in the long term; such factors including organizational effectiveness, industrial relations, and the capacity to produce quality products economically." (P. 220) If we say organizational instead of industrial, and realize that industrial relations include relationships with unions, but also professional associations, standard organizations, and the marketplace for workers, for inputs, and to our customers, then we have a good working definition. In psychological and social factors, we need to keep in mind the issue of cultural assumptions that we addressed in *Chapter 1: Quality Throughout History.* People may be unable to comprehend the approach we want to take. For example, when I heard about TQM using suggestions, I certainly did not envision the scale of the Japanese corporate suggestion management systems. I still have a hard time imagining an American company working effectively with suggestions on that scale.

We can then say that an organization is a system relating to other systems, contained within the larger systems of its industry, society, and the ecosphere, and meanwhile containing its own subsystems. From a systems perspective, we can define resistance as composed of from one to five elements.

- *Unhealthy resistance.* This results from poor communication and lack of willingness to create good communication and eliminate misunderstanding. This is covered by the word "politics" in the negative sense of the term, and includes the effects of all forms of dishonesty and avoidance.
- *Other priorities.* Any new initiative—when it is proposed, planned, or encountered by a new stakeholder—can appear to be in conflict with their priorities. We have to allow the stakeholder—person or representative of a subsystem—to identify their priorities, and then demonstrate that the quality program is either beneficial—or at least not harmful—to their goals, or is a worthwhile trade.
- *Constraints.* Any new initiative may exceed the limits of some part of the organization, or threaten to do so. We are also constrained by agreements such as union contracts, vendor contracts, and commitment to existing policies.
- *Inherent conservatism.* All lasting systems have a tendency towards homeostasis—a tendency to forcefully move towards their own pre-defined point of balance. If we are changing processes and results, we will be changing these set points. Some disruption is inevitable, so caution is appropriate. Sometimes that caution comes out as resistance; an effort to block or slow down what appears to be a dangerous process.
- *Inherent momentum.* Systems, like physical objects, tend to keep moving in the same direction that they already are moving. In a system, a successful change of direction requires not force, but encouragement for habit change.

Note that four out of five of the forms of resistance express completely legitimate perspectives, raising issues that we can and should take into account. The fifth—unhealthy resistance—is a natural part of being human. If we wish to work with those who resist our proposals, we must listen well, then help them educate themselves on the safety and value of our ideas.

Here are some common examples of types and sources of resistance:

- *The need to prove quarterly revenue.* In public companies, executives face serious consequences when quarterly financials are not good or not in line with expectations. This can block focus on long-term objectives, impose constraints, and make other
- *Faster, punchier options.* Other alternatives, such as mergers and acquisitions, may seem like an easier way to go than looking at—and dealing with—internal quality problems.
- *Quick fixes that sound credible.* If you know how difficult the problem is to solve, you know how expensive it will be. What do you do when someone claims they have a cheaper solution?
- *Vested interests.* Unions, managers, executive committees, and others will have a variety of interests—legitimate or otherwise—and we will need to address them proactively.
- *A belief that the solution is impossible.*
- *A lack of willingness to look at the problem.*
- *Resistance to personal change.* When we change processes, we don't only change the organization, we call on individuals to change as well. The self-image, as a self-sustaining system, naturally resists this. If we truly want to transform processes, we must support people in truly transforming as well. This can be looked at from various perspectives, such as emotional intelligence for business or the difficulty of changing habits and character.

Later in the chapter, we will discuss how to define the cultural issues and resistance that you are likely to face in your own company. In addition to addressing cultural limitations and resistance, you will also need to deal with two other issues. One is quite simple: You will have to show how your program and new processes are affordable at present, acceptable over the short- and medium-term, and beneficial to the company's longterm bottom line.

The other is that you will need to define quality—and especially what you include in a quality management program and what you leave out—clearly enough so that everyone understands it. That is, you'll have to succeed at a job Plato thought was impossible. See the suggestions in the section *A Quality Improvement Program* below.

Critical Success Factors

If we want to succeed where many companies have failed in the last hundred years, it would be good to know the critical factors for success. The most honest answer would keep this section very short: Deming already defined them for us. Go back to *Chapter 11: Total Quality Management* and read his 14 points. Study them. Form a discussion group about them. Update them. See how they apply to your organization, to every stakeholder and subsystem in and around your organization. Supplement the 14 points with other material from Deming, Crosby, and others. Deming worked to make quality management work, and I'm sure he would have had a shorter list if all 14 points weren't essential.

Let me pick one of Deming's points and show you what I mean by applying it to one subsystem of a company. My favorite is *#8: Drive out fear.* Since leadership is often best done by example, then encouraging others to follow, the first corporate subsystem I will look at is myself. I begin with a rigorous self-assessment. What fears do I have? How do they affect my work in general? How do they limit my ability to understand and implement this program? I break that part down step by step. Is fear keeping me from designing the program? Is fear a barrier to listening to others and including their resistance and their viewpoints into the plan? Is fear a barrier to my presenting the plan and advocating for it? And so forth.

If I find a fear, it's time for PDCA. Have you ever planned to eliminate a fear? If not, don't worry, others have. There are plenty of best practices for the elimination of fear. Which one you pick will be determined by your cultural preferences and openmindedness. If you want to limit yourself to popular but proven business methods of the last 20 years, try Stephen R. Covey's *The Seven Habits of Highly Effective People* or training in emotional intelligence for business. If you like a holistic, bodymind, New Age approach, try a ropes course or an Outward Bound program. If you want to go all the way outside the box, think about the Bible's guidance, "Perfect love casts out fear." Or turn to the methods of meditation from Buddhism, where awareness of all of our thoughts and feelings teaches us to transform them, cut out the roots of fear, and be liberated from confusion and suffering.

Did you hit any resistance when reading the last two paragraphs? It's good if you did. That was my interpretation of what it means to drive out fear, not yours. Feel free to throw all of it away. As Six Sigma teaches, challenge every question and every answer.

But don't throw away Deming without a close look. Seriously consider what he and other leaders in quality management have been trying to tell all of us for over 100 years. It will work if we listen, understand, make it our own, and then

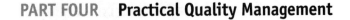
follow through on own convictions. If you want to overcome the challenges you will face in leading a quality management initiative, then make it truly your own, believe in it, and follow your convictions.

ALIGN YOUR Q

A Root Cause Analysis of Deming's 14 Points
I chose #8: *Drive out fear* as my example in this section for another reason, as well. I think that fear is a root cause of: lack of constancy; unwillingness to learn; and unwillingness to look at and accept the truth. As a result, it seems to me that fear is a root cause for many of Deming's 14 points. What do you think?

A Quality Improvement Program

In defining and presenting a quality improvement program, I would begin with these points:

- *Improved process for a better bottom line.* Improving the quality of processes improves product quality, reduces costs, and cuts cycle time.
- *Continuous improvement.* Our program will include a change of attitude and focus for everyone in the company, starting from the top down. Everyone will be supported in seeing how to make things better in their own work, for the customer, and for the company. This will lead to many low-cost initiatives—small changes with small benefits that add up to big results in cost performance, reduced cycle time, customer delight, and flexibility.
- *Executives set the big goals.* As an executive team, we will set big, hairy, audacious goals (BHAGs) for the company. These are achievable, but they demand innovation and thinking outside the box. They lead to highly profitable breakthroughs in process, products, and customer growth.

Here are the steps I would follow as a senior executive in any organization to implement a program for the improvement of quality management. My thanks to Malcolm Ryder, Principal and Research Director of Archestra, for brainstorming this with me.

1. *Float the idea and brainstorm it with the executive team.* If you're wrong, and a quality improvement program is not best for the company right now, you'll want someone to tell you right off. If you're right, you'll want the executive team to be able to say, "we came up with this great idea."

2. *Brainstorm even more.* If the basic idea is a go, then brainstorm the best name, sources of resistance and support, how to investigate standards and best practices, and how to deploy it internally.

3. *Develop the brainstorm results into a feasible—if audacious—strategic plan, then beat up on it.* Have the executive team create a plan that looks workable, then run it through program analysis, portfolio analysis, SWOT (strengths, weaknesses, opportunities, threats) evaluation, and anything else you can do to test the idea.

4. *Make a marketing plan and sell the idea.* Figure out how the program is likely to benefit each stakeholder group, and give them a straw man to shoot down. Then involve other leaders and stakeholder representatives the same way you brought in your own management team. Listen to them, and let them be convinced, and then sell you on your own idea.

5. *Create a team to lead the way, and give it full support.* Have a team build a plan that includes training, deployment, management, and feedback. Review it thoroughly, and show up for kickoff day.

6. *Stay on top of the program.* Make sure it is working well, and promote it. Assign resources to problem areas, and publicly reward successes.

7. *Listen and respond.* Each worker in the company will make this his or her own if you lead the way in appreciating every idea—good or bad—and in making decisions based on understandable, empirical, and fair evaluation.

8. *Keep it simple.* Some parts of this will be inherently complex, so it pays to keep everything as simple as it can be.

9. *Have the company make it their own, and keep truing the wheel.* The program must become an ongoing process of managed quality improvement.

EYE ON THE BALL

Make the Program Work
Here are some other key points in establishing a quality management improvement program.

- Quality Management always means bringing quality—something as large as life itself—under management, again and again. We have to realize that we are taking on a huge job, and we need to always be willing to start over.
- Quality always has two sides: an immeasurable side that we can sense, related to aesthetics, art, and design; and an engineering side that is definable, quantifiable, and measurable. We must always address both.

- A key step in quality management is to understand the principles of the cost of quality and, as a result, focus on preventing error from entering the system and eliminating it as soon as possible, instead of allowing errors to enter and struggling to eliminate them at the end.
- Effective quality management must also always have two sides: a human side, and a technical side.
- The human side of quality management must be both an individual and a team effort.
- The technical side of quality management—quality engineering—demands precise application of scientific methods of investigation.
- Both the human side and the engineering side can strive toward continuous improvement.
- We can improve quality through both innovations and incremental continuous improvement, but we always need continuous improvement to maintain quality and prevent backsliding.
- Continuous improvement will always run into roadblocks and distractions. The roadblocks will be cultural issues, where our growth takes us beyond where we are willing to go as individuals or as a team, and the distractions are the perpetual quick fixes of our society, such as short-term profit and dishonest representation of the situation.

Quality Management Self-Evaluation

Perhaps you don't need a completely new quality management program, or a full overhaul. Perhaps you just want to evaluate and improve what you already have. So, whether you are a senior executive or a manager with responsibility for quality, you'll want to know what you're doing well and what you could do better.

Here are some questions to ask yourself and your team:

- *How do we rate to Deming's 14 points?* Identify any discrepancies, and decide if they have significant consequences.
- *Does our program have full executive support?* Or is it too low down in the company to make process changes, or sidelined outside work processes?
- *Is our program visible to every employee, stakeholder, and customer?*
- *What is the history of our program?* Does that history imply any biases or limitations?
- *What are the strengths and weaknesses* of our program and our team?
- *Are we tracking our program?* Do we have metrics for the process and results of our quality management and quality improvement work?
- *Are we benchmarking?* Are we using those metrics to establish a baseline, and then using PDCA to improve.

- *Are we using best practices and standards well?* So much is available for free or at low cost, we should be using it. Assign someone to research what is out there for general fit, and make a practice of looking for solutions in existing methods when a problem is defined.
- *Does everyone have an eye on quality?* Maybe this should be first on the list.
- *Are people learning about quality and quality management?* People means everybody.
- *Are we recognizing and rewarding quality work and quality improvement?* Studies in business psychology affirm that the Japanese approach to rewards—prompt, small, and frequent—works best for everyone.

Quality Management for Managers and Workers

This section is for managers, supervisors, and workers. If you are in an organization that, overall, cares about quality, but has dropped the ball on quality, you may be able to pick up the Q-Ball for your team. Most North American managers have enough autonomy to at least support, encourage, and reward attention to quality problems. You may have to work very quietly, but you can work with your team, encourage attention and good quality work, and praise good work and suggestions for making the workplace better. If you want to do this, try to fit into the language and program that your company uses. Keep it small and personal. And, above all, be consistent. This approach will also work if you are a manager or department head with accountability for results and a reasonable degree of autonomy. If you are a project manager, see the next chapter.

As a worker, especially a worker in a quality-related job, you can make many of the ideas in this book your own. Perhaps you can even start your own Quality Circle for study and learning.

Conclusion: Quality—A Complete and Lasting Solution

We've seen many partial solutions to the quality management problem. What would a complete and lasting solution look like? Certainly, some companies—

in Japan and elsewhere—have one. And sometimes it is called TQM, sometimes ISO 9000, sometimes CMMI, sometimes Six Sigma. These companies have moved beyond standards and methodologies to lasting solutions. What makes the difference?

I would suggest that the bottom line when it comes to quality versus culture is that lasting quality, continuous improvement, and innovation in quality require a commitment to truthfulness—honesty and integrity. Quality is achieved by the definition and elimination of defect and errors. Let's call them mistakes and boo-boos. Why? Because, in most cultures and for most people, it is embarrassing to look at and talk about our mistakes. And, to do quality management, we have to see errors, be honest about what we see, and follow through with a solution. That's honesty and integrity—truthfulness.

In the fields of quality management and project management, confronting issues while working with people is a skill we learn and a character trait we develop. Most people are not willing to take that step because most business environments do not support it. For many people, it is easier to smooth things over or to walk away than it is to look in a mirror and see what improvements we need to make.

But for anyone willing to consider change, the option of quality—in process, in product, in profit, and in personal performance—is there. And, as both Deming and Imai affirm, the satisfaction—or even joy—of doing good work and solving problems together is worth the challenge of learning to be true.

In *Chapter 18: Practical Quality for Projects and Programs* we will see how quality management applies to a company's projects—those one-time endeavors to create products or services or make internal changes. Even if your company doesn't do a lot of projects, this will help because it will show you how to organize your quality improvement program—with quality!

Q-Ball Quiz for Chapter 17

1. Which of these is *not* true of resistance?
   (a) Resistance is entirely a negative thing.
   (b) Resistance has negative components, usually resulting from misunderstanding or dishonesty, but good components as well.
   (c) Resistance can show us risks to our plan.
   (d) Resistance is a sign that we should listen to the other person's perspective.

2. An organization's culture includes all of these *except*
   (a) industrial structure.
   (b) psychology.
   (c) agreements with unions and other parties.
   (d) current market position.

3. Which of these is *not* one of the three main points in the quality improvement program in this chapter?
   (a) Continuous improvement
   (b) Improved process for a better bottom line
   (c) Educate everyone
   (d) Executives set the big goals

4. Which of these is *not* a good reason to brainstorm a quality program with your executive team right up front?
   (a) To find out if it is really a good idea
   (b) So you can share the blame if it goes wrong
   (c) To let them make it their own
   (d) To identify sources of resistance

5. The most successful motivator is
   (a) fear.
   (b) big bonuses.
   (c) small cash bonuses given privately.
   (d) the opportunity to make a contribution with no barrier to its implementation, plus small, but frequent and personal, praise, encouragement, and recognition.

Practical Quality for Projects and Programs

They recognize the task before them as that of inducing each workman to use his best endeavors, his hardest work, all his traditional knowledge, his skill, his ingenuity, and his good-will in a word, his "initiative."
—Frederick Winslow Taylor, *Scientific Management*, 1911.

Total Quality Management guru Joseph M. Juran defined a project as "a problem scheduled for solution." This definition clearly links projects into the PDCA cycle, and implies that projects come in all sizes. I would add that there is another type of project as well—an opportunity scheduled for realization. By opportunity, I mean a customer's problem that we hope to get paid for solving, or a chance to offer something new or better, or to reach more customers and grow our business. The Project Management Institute (PMI) offers a more formal definition. For them, a project is "a temporary endeavor undertaken to create a unique product, service, or result." Programs are groups of projects and other activities that are tied together by a common goal, and portfolio management is the evaluation of a program's value, resources, and constraints. To learn more, go to *www.pmi.org* and take a look at their Guide to the Project Management Body of Knowledge (PMBOK® Guide), which is the ANSI standard for project management methodology.

Quality management and project management intersect at many levels. From the top down:

- If we want to overhaul quality and quality management at our company, the best way to do it is to create a program full of projects. The projects deploy the new methodology and improve quality, other program elements ensure that the quality methods and standards stick, and support continuous improvement so that improving quality becomes an unending process.
- Six Sigma quality management is organized around significant, rapid projects.
- Quality management is one of nine components—areas of knowledge—covered within project management in the PMBOK guide.
- Customer satisfaction is a key target of both quality management and project management.
- Project management contains tools, such as return on investment (ROI) analysis, that will help justify quality improvement programs, and risk management tools that will make your quality management program more likely to succeed.
- Every PDCA effort, no matter how small, can be viewed as a project, and applying project management methods can increase its value and its chances of success. Quality managers and workers doing PDCA will benefit from project management training.

So, if you're not a project manager, read this chapter to see how you can use projects to help with quality. If you are a project manager, then this chapter will show you how to apply the ideas in this book on your projects. Your relative autonomy to define and run the project means that you can create a quality teamwork environment for the project, even if that is not the norm for your company.

Project managers can increase the value of project results and increase the likelihood of project success by focusing on quality. Traditional project management focuses on three main elements: scope, time, and cost. The other six areas of knowledge—quality, risk, communications, human resources, procurement, and integration—are moved to a supporting role. However, if we evaluate this from a kaizen perspective, we will see that scope, time, and cost are the results (R) factors, while the other six areas are the process and people (P) factors. As we saw in Chapter 14: *Steady Improvement in Japan,* P factors are harder to measure, but often give earlier signs of trouble that can prevent problems in the R factors. The same is true in project management. Scope, time, and cost were identified as first because they were obvious, easy-to-measure success factors—or signs of disaster. But, for the most part, we make sure of their success by taking care of our P factors, by making sure that quality and the other five areas are brought and kept under management so that we can deliver the full scope of the project on time and under budget.

In project management, *scope* means what we are making, and all the work we will do to make it. Scope begins at a high level, but is clarified in greater and greater detail, creating a requirements specification. All of the issues about requirements specifications that we discussed in Chapter 3: *Defining Quality* apply to projects with their unique products or results, just as they do to manufacturing processes with their products. Perhaps they apply even more rigorously: If we are selling a product, we might get a chance to fix it later, but if we are delivering project results to a customer, it's a one-shot deal. In defining scope, quality management can be a big help. If we take this approach, quality management can be used to add value to the project while lowering cost.

Unfortunately, quality management is often sidelined in project management, and the reason it happens on projects is the same as the reason it happens in companies. If quality is seen only as conformance to technical requirements—such as standards and regulations—it is simply a pain to be dealt with. Only if customer requirements are included in the specification that we must adhere to through quality management does quality become important. So, we need to make sure that project quality management applies to all requirements—including the customer's.

When we do this, then scope definition—a formal project management process—and quality definition or requirements specification become one and the same thing. In scope definition, we ask the customer, "What are we making for you?" We ask over and over to get more and more detail. But, of course, the customer wants the item because it will add value. If we modify the question to ask, "What makes it good?" then we integrate scope definition and quality definition. With this approach, the goal of the project and the value of the project are more tightly linked.

Project managers are taught—appropriately—to avoid gold plating. Gold plating is adding things to the project that make the product better. Gold plating should be avoided for two reasons. First, our idea of value for the customer may be wrong. Second, even if the addition is greater value, adding it during the development stage of the project is ten times more costly and uses ten times as much effort as it would have in planning, due to the 1:10:100 rule. Many projects have failed because gold plating—even with the customer's permission or at the customer's request—took the project schedule and budget completely out of control.

Adding quality by asking "What adds value? What makes it good?" is not gold plating. The crucial difference is that we ask up front, at the beginning of the project. Anything we ask for will fit within the planning, or the "1:" part of the planning: development: delivery stages of the project, and will have much lower cost. In fact, doing this work may increase value while lowering total project cost. Getting a detailed requirements specification and understanding why each component is being built for the customer brings Crosby's Cost of Quality approach—which is included in the PMBOK guide—to our projects. We reduce project management costs, reduce quality management costs, and reduce life cycle maintenance costs for our customer. We also reduce risk by reducing the chances that we have missed or misunderstood a crucial stakeholder requirement. If a crucial requirement is missed or misunderstood, it will either be caught in the development stage, increasing project cost and creating delays, or it will be caught by the customer, which could lead to rejection of the project or to very high repair costs.

Now we have seen why good quality management should be integrated into every aspect of project management, with the most emphasis on early stages to take advantage of the 1:10:100 rule and optimize cost of quality. Let's turn our attention to how we do it.

Quality Processes for Projects

The PMBOK guide already includes performing quality planning, performing quality control, and performing quality assurance in its list of processes. The material on those topics in *Quality Management Demystified* exceeds those standards. The PMBOK guide openly encourages additional processes when they are found to be of value. Applying the approach of *Quality Management Demystified* to project management means adding two processes, one at the beginning, and one at the end.

PERFORMING QUALITY DEFINITION

It is clear that something must be measured before it can be managed. What is not always as clear is that it must be defined before it can be measured. Many things—such as time and cost—are defined for us already, and we have units of measure. But, as we've seen, quality is incredibly hard to define. And even when we define it, the quality—or error—in process has a very complicated relationship to the quality—or defects—in the product. So we need to define the quality of our product and our process very well, and do it early. We can do this by enhancing the project management process called scope specification to include specification of quality and assurance of value, as described above. Formally, we would add a new process to the PMI system called Defining Quality, and make it a core planning process in the quality management area. The output would be a scope specification that would be validated for containing answers to questions about quality and linked to the project value defined in the project charter. For more on quality definition, see Chapters 3 and 6.

DELIVERING QUALITY

The principles of customer delight apply to projects just as they apply to products and services. There are three differences between giving the customer what they asked for and delighting the customer. Remember, delighting the customer is not goldplating!

- *Expectations are more than specifications.* Customers tend to remember that they will get everything they asked for, even if we said "no." They also tend to dream. We can keep customer expectations under management by communicating with the customers periodically throughout the project, making sure that their expectations are in line with the specification we are working to deliver.
- *Getting it is not the same as knowing they got it.* I have seen a couple of projects fail simply because the customer never received what we delivered, or did not know that he received it. Following through with attention to the customer at the end ensures that they receive the item, check it for quality, and see that it meets expectations and delivers—or seems to indicate that it will deliver—value.
- *They may like our product, but not like us.* We need to make sure that our customers enjoy every interaction with the project team—that they receive prompt, professional, and friendly customer service at all times.

We can take care of the first point using project communications management. It also helps if we define the project lifecycle with stages and gates, as discussed in the Software Inspection section of Chapter 15: *The Capability for*

Quality: CMM and CMMI, and include customer review at the gate. The other two points can be handled by defining a new activity in the Project Close process group called Deliver Quality. In PMBOK terms, it would be a facilitating process in the quality management knowledge area. We would apply TQM methods of customer delight—see Chapters 4 and 6—to customer communications and product or service delivery.

Above, we have discussed how to integrate project quality management with project scope management, and how to improve project quality management itself. We have two more topics to address: integrating project quality management with all nine knowledge areas, and addressing quality management of the business and technical aspects of the project.

Quality Integrated into Other Knowledge Areas

The PMBOK defines about 40 processes that are part of project management. Quality, as we've seen, produces benefits by improving processes. We can enhance our chances of project success and enhance project value by applying best quality practices to all project management processes.

QUALITY IN PROJECT SCOPE, TIME, AND COST

In project management, Scope, Time, and Cost are considered the Iron Triangle—iron because it can't bend, and triangle because increasing the size of any one side of a triangle demands changes to one or both of the other sides. The lessons are: If the customer increases scope, they must put in more money, allow more time, or both; and, if the customer cuts the budget or delivery date, they must agree to a smaller scope by dropping some requirements.

This is true as long as our processes don't change. But as we've seen in Chapters 14 and 16, if we improve our processes with quality management, we can deliver more for less. Good quality management turns the iron triangle into gold.

We can do this by realizing that all elements of scope, plus the requirement to deliver within budget and on time, are customer requirements. As such, it is the responsibility of quality management to ensure that all project goals—scope, time, and cost—are met.

QUALITY AND PROJECT RISK MANAGEMENT

Quality management and risk management are parallel. Quality management defines error, brings error under management, and reduces error to acceptable levels—which may be zero defects. Risk management defines uncertainty—possible future events or outcomes—into specified risk, brings risk under management, and reduces risk to acceptable levels so that the project can continue to completion by successfully navigating all the risks. Aside from this parallel, quality management performs two main activities that reduce project risk. One is the careful definition of project scope and quality, which we have already discussed at length. This reduces the risks arising from delivering to specification, but not to expectations or needs. It also reduces the risk of later discovery of new essential requirements or later clarification of vague requirements. Every time we have to adjust requirements in the middle of the project, it costs ten times more than it would to have done it right the first time. A poor scope definition is a big risk factor. In my own experience, it's the largest cause of project failure. And quality definition will eliminate that risk.

The second is that we can apply quality assurance activities to every process on the project. We can assure—or even audit—the project to make sure that we are following the PMBOK standard, other standards such as CMMI, or internally defined best practices on each project, or on especially risky, costly, valuable, or high-profile projects. This will reduce the risk of project failure.

QUALITY FOR PROJECT HUMAN RESOURCES AND COMMUNICATIONS

The best way to apply quality management to project human resources and communications is to focus on the teamwork element. We can use the idea of the recipient of my work as my customer to facilitate a team approach to quality, as we discussed in Chapters 7 and 16. We can also use the idea—discussed under Software Inspection in Chapter 15: *The Capability for Quality: CMM and CMMI* and elsewhere—that there will always be error, so it is better to get support from our team members to find and eliminate it than it is to let the defect be discovered by the customer. When we use this inside the team, it is part of project human resources management. When we extend it beyond the team to include all stakeholders, including customers and vendors, it improves project human resources management.

QUALITY FOR PROJECT PROCUREMENT MANAGEMENT

Here, we can apply TQM methods of extending our own quality management to our vendors. One approach, discussed in Chapter 12: *Quality Standards—ISO 9000 and More,* is to require that our vendors have certification in an appropriate quality management standard, such as ISO 9000, one of its industry-specific variations, or CMMI. The benefits and limitations of that approach were discussed in Chapters 12 and 15. The other alternative is to simply incorporate—to the extent that it is feasible and that they will allow—customers as part of our own quality team. This was the original approach taken by TQM both in Japan and by the Big Three auto makers in the United States. I have done it successfully on a much smaller scale, incorporating as many as five vendors into a single project for a customer.

QUALITY FOR PROJECT INTEGRATION MANAGEMENT

Project Integration Management pulls all of the work of the other process areas under one umbrella. It also ensures that information from one area is properly communicated to other areas. For example, if there is a new requirement, that is a change in scope. The cost of that requirement should change the budget, the work required should be added to the schedule, potential problems in meeting the requirement become part of the risk plan, new skills needed are added to the human resource plan, and so forth. Quality management concepts are useful here because they help clarify how process (P) is related to requirements (R). In addition, many quality management tools, such as flow charts, Ishikawa (cause-and-effect) diagrams, and network diagrams can help us coordinate project work to achieve better integration. Lastly, if every team member is trained in quality awareness, it is easier for everyone to see the connections across processes and between processes and results. Once they see them, it is easier to address and discuss them in a quality team environment as well.

QUALITY FOR PROJECT QUALITY MANAGEMENT

Okay, I've got to admit that the title of this section sounds a bit redundant. But it is, nonetheless useful. There are two things we can do. First of all, we can ensure that we are doing the best quality management possible by researching applicable quality best practices as they apply to each unique project. Secondly, we can make sure that everyone on the project team is doing the quality work we planned to do through quality assurance on project quality management processes and, when necessary, quality audits of projects.

Quality Management for Projects at the Business and Technical Levels

Good project management is necessary, but not sufficient, for project success. For a project to succeed, it must be managed well at three levels—the business level, the project level, and the technical level. Let's look at the business and technical levels now.

QUALITY MANAGEMENT FOR THE BUSINESS LEVEL OF A PROJECT

The business level of a project is primarily concerned with four things: making sure that there is real value—usually thought of in terms of return on investment—in doing the project; making sure that the business supports the project as planned; making sure that product work has minimal negative impact on production work; and making sure that we manage the risks to the company that arise from project work.

Quality management can assist with each of these, as follows:

- *Value.* If the project is initially defined through PDCA, then it arises from a real problem in the company, making it more likely that project results—the solution—will be of value. Six Sigma project management—See Chapter 13: *Six Sigma*—has even stronger requirements for validating projects before they begin.
- *Support for the project.* Here, communication about value, and simply good communication—treating the guy upstairs as a customer—is essential.
- *Minimizing problems for production work.* If the production group— whether they are a source of workers, the customer of our project, or both—is treated as a customer, we will meet those needs.

- *Reducing corporate risk.* Corporate risk can arise when a project succeeds, but results have unexpected effects. Thorough investigation of
  project consequences—especially a description of the new process in
  place after a process change—reduces the risk that project success will
  make things worse than before we started. If a project fails and delivers
  no value, that is a financial loss for the company. Quality management
  reduces the risk to the company of project failure by making project failure less likely.

QUALITY MANAGEMENT FOR THE TECHNICAL LEVEL OF A PROJECT

I originally defined the five stages of quality management described in Chapter
6 by applying TQM principles to project management. I later realized that they
applied to quality work in the production world as well. So ensuring quality definition, planning, control, assurance, and delivery of the product components
and the whole product to the customer is the work of quality management at the
technical level. It is particularly valuable to apply a project lifecycle and then
insert quality control reviews as checkpoints at the end of each stage, as discussed under the topic of Software Inspection in Chapter 15: *The Capability for
Quality: CMM and CMMI.*

If the entire team is trained in PDCA, each project team member will be well
equipped to resolve problems that come up in the course of his or her work. If
the quality team approach is used, communication about problems will be simple, as well.

Quality Management for Programs and Portfolios

A program is a large endeavor spanning many years, composed of projects and
other activities. The classic example is the United States National Aeronautics
and Space Administration's (NASA) Apollo Project that took men to the moon.
The program included every launch of an Apollo rocket, including test launches,
trips to the moon, and trips that landed on the moon. Each launch was a project

with tightly scheduled activities and a specific set of goals. The program coordinated all the projects more loosely, so that, for example, when Apollo 13 didn't make it all the way to the moon, its payload of experiments could be distributed over the next several launches. The program also included non-project activities such as human resources and facilities maintenance.

If you are a program manager, you probably have sufficient leeway to create your own quality program using the method of your choice for the entire program. Selecting an approach and building in quality teamwork and methods from the very beginning—integrating the approach at all levels—will probably add value to your program in many ways. In a sense, you have an opportunity to do a total quality management implementation on a small scale.

Programs also need to be examined as portfolios. Portfolio management takes the tools used for investments—financial portfolios—and applies them to programs. It asks questions such as: Is the company getting the best possible return on investment from this program, rather than investing somewhere else? Is the company using all the resource of this program optimally? What are key constraints limiting the program's productivity? Questions like these can be answered with the help of quality management tools that map processes and organize issues. Many such tools can be found in *Six Sigma Demystified.*

Conclusion: Quality Management for Project Success

We can enhance the value of our project to our customer and our company by applying quality management methods throughout. Applying Cost of Quality methods doesn't just reduce the cost of quality management within the project, it can reduce the total cost of the project and also reduce the risk of project failure.

In addition, good project management practices are useful in the quality management field, as well. At a small level, any PDCA activity is more likely to succeed if defined as a project—a unique endeavor with a result. At a larger level, Six Sigma integrates project management and quality management. At the largest level, our effort to overhaul quality management within our organization is a project or program, and will benefit from project and program management methods.

Q-Ball Quiz for Chapter 18

1. Which of these is *not* one of the nine knowledge areas defined in the Project Management Guide to the Body of Knowledge?
   (a) Risk
   (b) Scope
   (c) Programs
   (d) Quality

2. Defining quality is closely linked to
   (a) defining scope.
   (b) defining risk.
   (c) defining cost.
   (d) defining communications.

3. Which quality management method most explicitly mentions projects?
   (a) Six Sigma
   (b) Gemba Kaizen
   (c) ISO 9000
   (d) Cost of Quality

4. Which of these is *not* one of the three levels of a project that must be managed to ensure success?
   (a) Project
   (b) Quality
   (c) Business
   (d) Technical

5. You are asked to explain why so much effort is being put into defining requirements—planning a project—and you are not just getting to work and getting it done. Which quality theory would support the extensive planning effort?
   (a) Total Quality Management
   (b) Six Sigma
   (c) Quality control
   (d) Cost of Quality

19

Global Quality in the 21st Century

We can see our forests vanishing, our water-powers going to waste, our soil being carried by floods into the sea; and the end of our coal and our iron is in sight. But our larger wastes of human effort, which go on every day through such of our acts as are blundering, ill-directed, or inefficient, and which Mr. Roosevelt refers to as a lack of "national efficiency," are less visible, less tangible, and are but vaguely appreciated.
—Frederick Winslow Taylor, *Scientific Management*, 1911.

Quality is a key issue for the entire planet. A focus on quality could have a profound effect on the nature of the global society we are creating, on the war on terror, and on the sustainability of earth's ecosystem for many generations. And, as my colleague Malcolm Ryder, Principal and Research Director of Archestra puts it, "high-quality companies have the best opportunity to be the ones that solve the problems we care about the most."

In this chapter, I am writing to all of us as people and as present or future leaders in business or other areas of society. In 1911, Taylor wrote, "It is hoped, however, that it will be clear to other readers that the same principles can be applied with equal force to all social activities: to the management of our homes; the management of our farms; the management of the business of our tradesmen, large and small; of our churches, our philanthropic institutions, our universities, and our governmental departments." I echo that sentiment here—wherever we are on the globe, and whatever roles we have in our lives, we, as people, are responsible for the quality of our lives and society.

Quality management call us to ask these basic questions:

- *Do our goals add value for* everyone *involved?* If we achieve our goals will—as Taylor put it—the company, the employees, and society all benefit? More specifically, will each group within society benefit? I propose that true and lasting value is value for all, not a win for some and a loss for others. If we play the win-lose game, we ultimately lose, because today's winners will be tomorrow's losers. But if we all win, we win for a longer time. And the notion of "all" can no longer be limited to people. As American President Theodore Roosevelt foresaw—echoed by Taylor in the quote that opens this chapter—"all" now includes all species and our natural resources.

- *Are we willing to look at our own processes?* At a business level, this means observing and measuring operational processes, so that we are ready to apply either continuous improvement—kaizen—or an innovation, whichever is needed. At a personal level, this is about self-awareness and introspection—the basis of character development and the basis of the development of emotional intelligence.

- *Are we willing to change our processes so that we can meet our goals?* To become effective at adding value, we must be willing to change what we do. As people, we will need to think and act differently. As businesses, we will need to change the way we work by developing and adopting best practices.

- *Are we willing to address every issue and make any significant change, large or small, based on the evidence?* This is the essence of the PDCA cycle applied to every gap and source of waste in our process.

- *Are we willing to work with the information we have?* Systems possess a tremendous amount of information. Most of it goes to waste. Quality management leads to performance management and knowledge management because it calls us to put our knowledge about our processes and results to good use. The next phase of quality management might be to ask the question: What knowledge do we have that we are letting lie around? How could we use it?

In the next two sections, *Quality in and out of the closet* and *From national dominance to national servant leadership,* I will be addressing U.S. business and its effects on global society. In the third section *Consumers, Customers, Employees, and People,* I bring the discussion back to all of us: to each reader, each person, each potential leader in any country. I ask you to join this discussion at two levels: global vision in *quality and global society* and local action— what you can do in your own business and your own life, in *creating sustainable, growing quality* I invite you to dream and plan with me.

Quality In and Out of the Closet

A company that is truly high quality must be able to sustain quality steadily over a long time, even when there are profound or unpredictable changes in the social or business context or in internal systems. Dr. Imai proposes a recipe of small amounts of innovation, lots of continuous improvement—kaizen—and lots of good work. Taylor proposed much the same in 1911, and said that such companies would themselves benefit, the workers would benefit, and society would benefit.

In the early 1980s, American business faltered, and, at first, blamed issues outside of its control such as the newer industrial infrastructure in Japan and Germany. Then we embraced quality management, decided it was a fad and got tired of it, downsized and re-engineered to save the bottom line, sometimes with little concern for quality. Our talent for innovation rescued us when we launched the Internet and rode the dot-com bubble while inventing huge advances in consumer electronics. All in all, we had a good couple of decades. Now, things are slipping again, and we are again blaming things outside our control—this time the outsourcing of jobs to India and Asia, and the low salaries there. Perhaps it is time to, once again, turn introspective and ask why we Americans are not doing as well as we might. Why have we created CMMI, but now lag behind the People's Republic of China in making it a national standard? Is it possible that our belief in innovation has gone too far, and that we can only improve in a rapidly expanding economy, we can't remain stable in times of slow growth?

Is it possible that we are actually addicted to innovation, to quick-fix solutions, because we are unwilling to look at the underlying problems? Consultants, rather than recognizing that they have simply enhanced and repackaged the work of earlier generations, need to declare their professional predecessors failures and claim to provide the solutions. Companies call past programs failures and look for new solutions. Perhaps, though, if we have the same old problems, we don't need new solutions.

As long as we keep running after new solutions from outside and relying on innovation, American quality is likely to yo-yo: up and down. If we want to stabilize, we will need to address our cultural issues the way that many Japanese companies have through gemba kaizen. On the plus side, this is certainly possible, as there are high quality companies of all sizes in the United States and around the world. A good place to start to find them is "*Fortune*" magazine's list of the 100 best companies to work for. Two of those, Continental Airlines and Whole Foods Markets, have impressive quality throughout the company. Other companies, such as Xerox and UPS, built quality systems back in the 1980s that have endured. On the other hand, the diversity of culture and—more and more—language—in the United States makes the creation of a unified corporate culture with common values more difficult. Collins and Porras's survey of visionary companies, *Built to Last: Successful Habits of Visionary Companies,* while not directly about quality, is very much about the ability to sustain excellent corporate culture and leadership, and includes practical tools.

From National Dominance to National Servant Leadership

If I could speak to the Board of Directors of Corporate America, I would share these thoughts. It is probably good that there is no such board. Our independence within our society is a strength. In this section, I will talk about what America—or any country—might do. Later on, we'll get down to nuts and bolts and look at what each of us—in our companies and our lives—can do.

If we can create more quality companies—either by improving our major corporations or by growing a new generation of quality startups—what should we focus on? In an increasingly global society, each country must find its niche based on its situation and culture. In the past, America has held a place of global economic leadership through the financial power of its natural resources, the value brought by its educational system, and its cultural gift for innovation. As American natural resources dwindle, China is beginning to tap into vast untouched reserves. Both China and India are surpassing us in education for business and technical work, and their salaries for professionals will be lower than U.S. salaries for years to come. Perhaps our gift for innovation is our only remaining edge.

Where, then, should we focus this innovation? Innovation is good for solving problems that have not been solved before, and for solving problems on a large scale that have only been solved on a small scale. What are the world's problems for which we could develop solutions?

QUALITY, ECONOMICS, AND THE ENVIRONMENT

These days, everyone is talking about China as a global powerhouse. And, using economic indicators, they are right. But economic results are created by business activities, and those are created by—and affect—society. They are also created from—and affect—the environment. In any system, nothing can get larger forever. And large things tend to collapse. As a culture, China is making choices—particularly in relation to the mass migration of people into cities and the pollution of the environment—remarkably similar to choices that caused England to rise and fall as a global power, and caused the United States to rise and now to begin to slip. It is highly likely that, in 40 or 50 years, China will pay a lot of money for innovative solutions to massive environmental problems such as pollution and even global warming. A technological solution for global warming? Changing the planet's temperature! That sounds absurd. But one has already been proposed. On the down side, it was proposed in a science fiction magazine—*Analog Science Fiction and Fact.* On the plus side, many proposals that originated in science fiction have now become real products of American ingenuity.

Think of it—North America could begin a continental Quality of Life initiative, addressing process quality, quality for business and profit, and quality of life through all of our social institutions. If you read all of the literature on quality and excellence from the 1980s and 1990s, you'll see that we almost got there. TQM infected more than just business. Some governmental agencies were transformed. Universities, schools for children, and hospitals all adopted TQM. TQM was changing all aspects of society.

What does TQM have to say about globalization? Globalization—a transformation to an information and economic system where information, products, and services find the barriers of national boundaries and distances are much less relevant than they used to be—is unavoidable, and we are seeking the best ways to do it. We are in a difficult situation that will require change—and loss for some—in the process. Whether it leads to benefit for all soon, or not until much later, may depend on whether we play globalization as a win-win-for-all game, or as a win-lose game.

The term "globalization" had two meanings in the business literature of the United States in the late 1990s. One, offered by leaders such as Stephen R. Covey of *Seven Habits* and Gordon Bethune, then CEO of Continental Airlines, who were both strongly influenced by the TQM movement, offered a win-win-for-all vision of peace and economic stability with cultural equality. The other—which right now in 2005 seems to have more influence—operates more from fear and an effort to control, with a win-lose mentality.

SERVANT LEADERSHIP

There is a longstanding alternative model—sometimes the dominant model—in American culture, called servant leadership. John Bogle, inventor of the mutual fund and founder of the highly successful Vanguard Group proposed Servant Leadership as a model for all of us. I think that servant leadership—problem-solving for the good of all people—is what American culture offers when Deming's 8th point—the elimination of fear—is present. We take on audacious challenges and succeed for the benefit of all. If we are willing to turn inside ourselves and eliminate fear, then we can solve our own problems. And the problems we have today—industrial pollution and excessive urbanization, for example—will be China's problems in a few decades.

Are we willing to experiment on ourselves and solve our own problems? Right now, as we are the only major nation denying the consensus on global warming, it seems unlikely. However, throughout the 20th century, the United States and other nations made rapid and unexpected political and social changes for the good. Maybe we will again. Isn't that up to us?

Consumers, Customers, Employees, and People

Up to us. I don't mean up to American business. In fact, this issue is not limited to America, and it isn't limited to business.

Let's take a closer look at the closing paragraph of Taylor's *Scientific Management:* "It is hoped, however, that it will be clear to other readers that the same principles can be applied with equal force to all social activities: to the management of our homes; the management of our farms; the management of the business of our tradesmen, large and small; of our churches, our philanthropic institutions, our universities, and our governmental departments." It seems to me that all of these organizations have two things in common—they are run by people, and they change the conditions in which we live.

Quality is as large as life, and our quality of life is at risk. With global warming causing hundreds of thousands of deaths a year, our life itself is at risk.

We—the people—created these problems, and we can solve them. That is the basic gift that America offers the world in three simple words, "We, the people."

But we've stopped seeing ourselves as people. We're consumers of business products. We're employees of companies. We're even users of computers and drugs. All too often, we've come too close to forgetting to be people, and forgetting the deeper values that make for a good life. We need to restore our vision of ourselves as people, and define what makes life good in a fresh and lasting way. When we do this, we change what we mean by the value of life. Value for our own life—informed self-interest—value for those around us—altruism—and value for the planet—global vision—take on a new relationship to one another.

I'm not talking about "the good life" or value in a superficial sense. I'm talking about the deep reward of living a life dedicated to the good, of living a valuable life, of living a life that adds value to all life on the planet. Many people around the world do this. Some work in business, others in government, education, or the not-for-profit realm. Many of us do it in our homes. Don't children warm our hearts and enrich our lives each day, just playing and learning? Three hundred years before the publication of this book, Benjamin Franklin was born, and he lived a life dedicated to the public good. His innovations included free public libraries and fire departments, among many others. He developed the system of time zones that now synchronizes the world's clocks. Americans could do well by living up to the heritage of being born in a nation that had Benjamin Franklin as one of our founding fathers.

Perhaps it is more important to re-envision ourselves than to re-envision our nation and the world. A global vision is good. I would expand French sociologist René Dubos's famous epigram "Think globally, act locally," to "Dream globally, act locally." The bottom line is, given life and the rich Earth that sustains us, we create our lives within this system. We reshape the system into civilization, and then live in that civilization, including business, religion, politics, social activities, home life, and all the rest. And what we create, we can change.

Plan, do, check, act. Planning should always include a global perspective, so that we can be sure that what we do is good for all. Doing, checking, and acting are always local work. Though, if our influence grows, our local doing might lead to local doing across the entire globe. Shewhart worked locally for AT&T. Deming and others carried that work around the world. I encourage you to work locally, even if your business, like mine, has just a few people in it. I encourage you to work locally inside yourself, and then to do it in whatever area of life or culture is most relevant to you. E. M. Forster, the English novelist, in his *Aspects of the Novel,* proposed that value does not operate inside the timed events of our daily lives. Instead, it operates outside the step-by-step "what happens next" inserting new meaning and new value. How can you bring new meaning and new value to your life and the people around you? Only you can answer that question, because only you are in the gemba—the real place of your life, on the spot, where the action is.

We are all on this globe together, in the gemba of planet earth. Let's dream together. What would a life of quality—a life free of fear—look like for all seven billion of us? What would business, education, social, and home life be like? Dare to dream globally. Then act locally. Let's each begin with ourselves, and then help one another.

Quality and Global Society

Would a global society of quality have war? Would it have starvation? Would it have fear and oppression? Let me be clear. Let's look deeper than the obvious answer that says, no, a world free of fear would not have these. Perhaps, somehow, we need them. Would we have refused the Civil War if that meant that slavery would not have to end? If everyone was safe and well fed, what would happen to motivation? England and Japan have faced that challenge in different ways.

Historical challenges lead to large, practical philosophical questions: How can we resolve the uneasy balance between stability and freedom? How would we ensure health and freedom from poverty while still motivating people to meaningful work? How would we handle the inevitable fluctuations of the economy? How would we handle the troubled people who, in their fear, trouble others?

I will not criticize those who think that war is necessary, or who prepare for war. But war is very expensive in so many ways. To a quality engineer, that means that war probably addresses symptoms that have already been produced. And if we can address the causes of war, building quality into people's lives so that war is not needed, then the solution will cost 100 times less.

Are these questions too large for a book on quality management? I don't think so. Quality is as large as life itself. If we are going to bring quality under management, we are going to give ourselves well-managed—orderly yet free—lives. As we noted in Chapter 1: *Quality Throughout History,* the opportunity to do business and improve quality in business depends on having a stable society. How can we use our businesses to promote a healthy, stable society?

Okay, you say, it fits the topic of quality. But what about management? Isn't this book about business?

I reply—how has business changed our lives and the world in the last 11,000 years since the Sphinx was built? How do we want business to change our lives in the future? How about tomorrow? How about today? Each day, we have a chance to move in the right direction. Dream globally, act locally.

Creating Sustainable, Growing Quality

I am not asking you to run out and solve the world's problems. If your life and your business are like mine, then you have your own share of problems to solve already.

Our future will not be created so much by grand visions as by how we address the problems in front of us, and what vision guides us as we do.

One of the most important changes in the American economy is the growth of small businesses. It is a huge cultural change with far-reaching impact on how America lives, and on how we produce innovation. Small business innovations are picked up by large businesses around the globe quite quickly. And the small-business approach is changing the way people work in large businesses as large companies reorganize the workplace—creating more flexible environments, more effective small teams, and communications systems that allow people to work from home or on the road—to compete with the benefits that small businesses offer their teams.

Taylor, Deming, Imai, and Crosby all went from large business environments to small business environments to get the word out about their vision. Wherever you are, coming up with new solutions—whether the great leaps of innovation America prizes, or the small steps of Japan's kaizen—opens the door to making a bigger difference.

So, please, pick up any one of your problems. Then:

- *Plan.* Define the problem clearly. Research to find out if anyone has already solved a similar problem. (The Internet is such a gift when it comes to finding best practices.) Brainstorm for innovation. Use the practical common-sense approach of kaizen. Think inside and outside the box, and come up with the best solution you can.

- *Do.* Build a team and, cooperatively, make it happen.
- *Check.* Did your problem get solved? Adjust as needed.
- *Act.* Make the solution permanent. But then go further—share the solution with the world. Open the door so that others can make the solution global.

Conclusion: Quality and Our Future

Quality management puts the fundamental connection between process and results in our hands. We can observe what we have. If we don't like it—PDCA. If we like it—continuous improvement. Working within our own lives—our circle of control—at home and in business—we can make a difference. As we make a difference, we impress people, and our circle of influence grows. Vaclav Havel, Aung San Su Kyii, Nelson Mandela, and Anwar Sadat were all once in prison. Each one became leader of his or her own nation—the nation that imprisoned him or her. Later, each won the Nobel Peace Prize for contribution to humanity. How did they do it? I don't completely know.

Clearly, each of them had strong traits of quality that carried the day. They found those traits within themselves. Anwar Sadat said he learned that strength in prison. Cultivating their strengths, all four of them produced their triumphs, step by step.

I do know that you and I have the same intrinsic gifts. Let's be aware. Let's define quality for ourselves in a way that benefits others. And then let's work together.

Tomorrow, you will get up, and, if it's a weekday, go to work. What quality will you find? What quality will you create?

Somehow, a multiple-choice quiz doesn't seem right after questions like these. Instead, send me your thoughts. You can find me in gemba, where the action is, at *www.qualitytechnology.com*.

Final Exam

1. Quality control has these two meanings:
   (a) Broad—all checking, and narrow—statistical quality control.
   (b) Broad—all QA, and narrow—statistical quality control.
   (c) Statistical quality control and statistical quality control plus process improvement.
   (d) PDCA and DMAIC.

2. An executive of a Six Sigma company wants to ensure business value, and also encourage innovation. The best tool to use is
   (a) outside-the-box thinking.
   (b) setting a base goal plus a stretch goal for the project.
   (c) setting a moderate sigma level as a base goal and high sigma level for the stretch goal.
   (d) setting a base goal based on the voice of the customer, and a stretch goal at a high sigma level.

3. All of these are true of unhealthy resistance *except* it
   (a) results from poor communication and lack of willingness to create good communication and eliminate misunderstanding.
   (b) includes politics in the negative sense of the term.
   (c) can be totally eliminated by creating a hassle-free environment.
   (d) includes the effects of all forms of dishonesty and avoidance.

4. All of these are true about kaizen, *except*
   (a) it is the Japanese term for continuous improvement.
   (b) it is part of TQM.
   (c) the literal translation is change-good.
   (d) it is part of Six Sigma.

5. Which of these is not one of Deming's 14 points?
   (a) Eliminate fear.
   (b) Break down barriers between staff areas.
   (c) Eliminate slogans, exhortations, and targets for the work force.
   (d) Eliminate barriers to vertical communication.

6. An independent company that, for a fee, audits you to see if you meet the ISO 9000 standard is called a
   (a) validation agent.
   (b) registrar.
   (c) ISO 9000 auditor.
   (d) certification agent.

7. Software Inspection
   (a) applied Total Quality Management to the process of software development.
   (b) applied the Zero Defect Movement to the process of software development.
   (c) applied Six Sigma to the process of software development.
   (d) applied continuous improvement to the process of software development.

8. In statistics, estimation means
   (a) making statements about the future based on the past and present.
   (b) making statements about cost or time, not quality.
   (c) approximation.
   (d) making statements about the population based on statistics from the sample.

9. Scientific Management
   (a) is the name for the predecessor of quality management.
   (b) is a tool used in Six Sigma.
   (c) was developed by John Smeaton.
   (d) is the English translation of the Japanese gemba kaizen.

10. In auditing, an effect is
    (a) the gap between condition and criteria, and its consequences.
    (b) the result of a defect.
    (c) the result of an error in process.
    (d) a measured result.

11. Which of the following factors is *not* involved in Total Quality Management?
    (a) Quality teams
    (b) Control charts
    (c) DMAIC
    (d) Executive leadership

12. Scientific Management is
    (a) a cutting edge movement in Quality Management.
    (b) a product of *Gemba Kaizen.*
    (c) another name for TQM.
    (d) the predecessor of all quality management.

13. Which of these is *not* a major barrier to quality improvement programs?
    (a) Executive management often doesn't make it easy for managers and workers to change processes.
    (b) Executive management doesn't think quality and productivity must be at odds with one another.
    (c) Quality improvement programs can't always show the quarterly gains executive management feels pressured to provide.
    (d) Total Quality Management is only proven to work in Japan.

14. Quality management uses these elements from the field of science *except*
    (a) the scientific method.
    (b) an empirical approach to problem solving.
    (c) the search for understanding how the world works.
    (d) experimental design.

15. In auditing, the reason for the good or poor performance is called a
    (a) cause.
    (b) root cause.
    (c) measurement.
    (d) effect.

16. Zero defects
    (a) was first achieved by the developers of the software for the space shuttle.
    (b) was first achieved by Six Sigma engineers in the 1990s.
    (c) was first achieved by a company building missiles for the U.S. government in the 1960s.
    (d) has never been achieved.

17. All of these are true of quality planning *except*
    (a) quality planning is a focus in ISO 9000.
    (b) quality planning defines both QC and QA, and more.
    (c) quality planning increases the total cost of quality.
    (d) quality planning is a newer process than QC or QA.

18. The five S's are
    (a) Six Sigma Saves Software Systems.
    (b) Sort, Straighten, Scrub, Systematize, and Standardize.
    (c) Standardization Systematizes Saving Support Services.
    (d) Standardize, Systematize, Straighten, Sort, Support.

19. These are all legitimate reasons to resist a new initiative, *except*
    (a) constraints.
    (b) inherent conservatism.
    (c) inherent momentum.
    (d) All of these answers, a, b, and c, are legitimate reasons to resist a new initiative.

20. The Deming curve
    (a) illustrates the value over time realized by TQM.
    (b) summarizes statistical quality control results.
    (c) summarizes the results of PDCA.
    (d) is a made-up term.

21. Which of these is *not* true of Deming's 14 points?
    (a) They apply only to Total Quality Management.
    (b) They help with a root cause analysis of the problems in any quality management initiative, no matter what method is used.
    (c) Some are phrased in the language of the time, but all point to universal principles.
    (d) They are a comprehensive list key points for a TQM program.

22. Which of these terms is *not* used in ISO 9000?
    (a) Registration
    (b) Document
    (c) Certification
    (d) Record

23. Philip B. Crosby is associated with all of these *except*
    (a) Total Quality Management.
    (b) Cost of Quality.
    (c) Do It Right the First Time.
    (d) Hassle-Free Management.

24. Work Breakdown Structuring is all of these *except*
    (a) a way of defining the scope of work in detail.
    (b) a tool used in Six Sigma.
    (c) a very important tool used in project management.
    (d) a tool often used in continuous improvement.

25. All of the following are true about the 1.5 sigma shift *except*
    (a) it was measured at Motorola, then used widely as a rule of thumb without retesting.
    (b) it is assumed in the most common definition of Six Sigma.
    (c) GE published a validation of Motorola's discovery.
    (d) many Six Sigma engineers don't fully understand it.

26. A comprehensive sample is
    (a) as much of the total population as we can get.
    (b) a large random sample.
    (c) an unbiased random sample.
    (d) the same as the population.

27. A unique contribution of Gemba Kaizen to quality management is
    (a) the suggestion system.
    (b) control charts.
    (c) a focus on innovation.
    (d) Pareto optimization.

28. In experimentation, a process that is not changed, and compared to a process that is intentionally changed, is called an
    (a) experimental group.
    (b) experimental baseline.
    (c) experimental benchmark.
    (d) experimental control.

29. Process improvement includes all of these *except*
    (a) reducing cycle time for each process.
    (b) consistently replacing inspection with quality control.
    (c) eliminating processes that don't add value.
    (d) improving the quality of output from each process.

30. In quality management, we should work to improve the quality of all of these *except*
    (a) our workforce.
    (b) our products.
    (c) our tools and processes.
    (d) our competitor's product.

31. Quality assurance includes all of these *except*
    (a) processes to evaluate and improve processes.
    (b) improved communication about problems throughout the SIPOC chain.
    (c) designing in quality.
    (d) statistical process control.

32. A systems breakdown diagram
    (a) shows points of failure in a system.
    (b) breaks down a system into its component processes, called sub-systems.
    (c) is a Six Sigma variation of a work breakdown structure.
    (d) is a made-up term.

33. You are a new senior executive for a company. You call your quality assurance team together and ask what your predecessor did wrong. They tell you that he never allowed them to really change processes, they could only try to patch things up after the fact. Which of these approaches addresses that issue most directly?
    (a) Six Sigma
    (b) Hassle-free management
    (c) The voice of the customer
    (d) Continuous improvement

34. The quality of people being able to do more together than we could do separately is called
    (a) entergy.
    (b) co-creativity.
    (c) synergy.
    (d) balanced optimism.

35. The two best kind of samples to use are
    (a) comprehensive and random.
    (b) random and stratified.
    (c) stratified and convenience.
    (d) population and random.

36. Checking includes all of these *except*
    (a) statistical process control.
    (b) review.
    (c) inspection.
    (d) designing in quality.

37. Flow production is all of these *except*
    (a) where workstations are arranged in order of work flow.
    (b) part of just-in-time manufacturing.
    (c) a technique for reducing waste.
    (d) a technique often applied in Six Sigma.

38. Which of these most accurately describes the development of TQM?
    (a) Deming brought TQM to Japan, Japanese business made it work.
    (b) Deming was the only outside influence—the Japanese did all the rest.
    (c) TQM was the result of a nationwide effort in Japan with major contributions by Japanese industrial leaders, Japanese management, Japanese workers, and others as well.
    (d) TQM failed after Deming left, and the Japanese moved on to Kaizen.

39. Impact Finding is a term from
    (a) Six Sigma.
    (b) auditing.
    (c) software inspection.
    (d) ISO 9000.

40. Constraints on a new initiative include all of these *except*
    (a) limits of the organization, such as available funding or staff, that might be exceeded.
    (b) union and vendor contracts.
    (c) commitment to existing policies.
    (d) the limitations of innovation.

41. Egoless programming
    (a) is supported by the idea that everyone makes mistakes, and we'd rather have our team find them than have the customer find them.
    (b) is not part of formal software inspection.
    (c) requires extensive training.
    (d) is an alternative to flowcharting.

42. Which is the best way to finish the following sentence: To the extent that a business does not do quality management,
    (a) it will save money by avoiding useless rah-rah programs.
    (b) it will get away with it as long as the economy is good.
    (c) it can make up for it in advertising.
    (d) the company and its customers will have to deal with the consequences of the errors the company did not resolve.

43. Six Sigma short-term projects have all of these advantages *except*
    (a) they give more focus than an ongoing quality program without specific goals.
    (b) they can be driven entirely by internal and customer requirements, unlike CMMI certification.
    (c) they are an effective approach to continuous improvement.
    (d) they tend to foster innovation.

44. You have embarked on a quality program for the past three years. By all measures except one, it appears to be working. The only change is that your company's share of the market isn't growing. All of these could be factors, *except*
    (a) your competitors are improving quality, just as you are.
    (b) your marketing team isn't getting the word out about your new quality, so customers don't know about it yet.
    (c) the industry's whole market is shrinking.
    (d) you haven't extended the program to the customer through customer delight yet.

45. All of these are true of pull production *except*
    (a) it was developed by SONY.
    (b) it is only likely to work after process is optimized and standardized.
    (c) it is part of lean manufacturing.
    (d) each workstation makes only enough output to supply input as needed for the next step.

46. All of the following are true of the cost of quality *except*
    (a) You can be certified as a cost-of-quality auditor or engineer.
    (b) The cost of quality can be used to validate any quality management initiative, large or small.
    (c) The cost-of-quality concept is a good way to sell your executives on an effort to improve the quality of business processes.
    (d) The cost of quality is recognized in the Project Management Body of Knowledge.

47. Inherent conservatism
    (a) is a result of a lasting system's tendency towards homeostasis.
    (b) is a result of stagnation.
    (c) is a barrier to continuous improvement, but not to innovation.
    (d) is a barrier to innovation, but not to continuous improvement.

48. Another term for an ISO 9000 document is
    (a) a quality plan.
    (b) a standard operating procedure (SOP).
    (c) a record set.
    (d) a quality control chart.

49. Which tool is best for prioritizing cause-effect pairs?
    (a) The cause-and-effect diagram
    (b) The Pareto chart
    (c) The histogram
    (d) The flow chart

50. Which of these statements is most true with regard to the idea of zero defects?
    (a) Zero defects should always be our goal.
    (b) Zero defects is only appropriate when human life is on the line, or when the cost of resolving the defect later is huge.
    (c) Zero defects is most important when human life is on the line, but can be applied elsewhere as well.
    (d) Zero defects is impossible to attain.

51. SOP stands for
    (a) Standard Operating Procedure.
    (b) Statistically Optimized Process.
    (c) Same Old Problem.
    (d) Same Old Planet.

52. The internal auditor's attitude towards a management response in value-added auditing is
    (a) the auditor should recommend to management what the response to a finding should be.
    (b) an audit report only reports findings, the management response should come later.
    (c) the auditor should focus on findings, but help management develop its response, and include the management response in the audit report.
    (d) once management has defined the management response, the auditor must explain discrepancies between the finding and the management response.

53. Total Quality Management clearly demonstrated that
    (a) process improvement results in both improved product quality and increased productivity.
    (b) statistical quality control is always better than inspection.
    (c) inspection should be replaced by TQM.
    (d) we can improve quality, but the conflict between production quantity and product quality cannot be resolved.

54. Overall, quality management
    (a) is aimed at the customer, and also improves the bottom line,
    (b) does what is best for the bottom line, even if it means a dissatisfied customer,
    (c) is good for our customers, but cannot be made reliably good for the bottom line,
    (d) has no effect on the bottom line

55. The relationship between ISO 9000 documents and records is
    (a) documents tell you what to do, records show that you did it.
    (b) records tell you what to do, documents show that you did it.
    (c) you must have either a record or a document of each test.
    (d) documents are for QA, records are for QC.

56. In setting an acceptable level of error, all of these is true *except*
    (a) zero defects is an acceptable goal, but not the only alternative.
    (b) in allowing some defects, we run the risk of actually planning to include defects.
    (c) zero defects is impossible.
    (d) the customer's preference automatically determines the acceptable level of error.

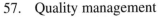

57. Quality management
    (a) can only be implemented through an intensive, structured program.
    (b) can be implemented incrementally, or though an intensive program, but a good plan is essential.
    (c) will be beneficial, even if done without strategic planning.
    (d) should be entirely guided by outside consultants, because their expertise produces better results than ideas from executives and managers.

58. Review
    (a) is for documents, not products.
    (b) is more expensive than testing.
    (c) is more rigorous than inspection.
    (d) eliminates the need for testing.

59. ISO 9000 requires all of these *except*
    (a) that you have a quality system (QS) in place.
    (b) a document describing how each procedure is done.
    (c) a set of records showing that procedures were actually followed.
    (d) a method of statistical quality control.

60. Just-in-Time (lean) manufacturing insures all of these *except*
    (a) at every stage in the production process the inputs necessary for the next stage are available, but without excess inventory.
    (b) items are manufactured only after they are ordered to prevent excess inventory.
    (c) the manufacturing process is streamlined.
    (d) best practices from the industry are in place.

61. Buddy programming is all of these *except*
    (a) an innovative approach that actually has two people working simultaneously at one computer screen to write code.
    (b) an inexpensive way to gain a degree of independence in checking.
    (c) less expensive—but also less capable of finding errors—than complete software inspection.
    (d) incompatible with software inspection.

62. Which of these is a problem with Six Sigma?
    (a) There is a high cost of training Six Sigma black belts.
    (b) Six Sigma focuses on innovation, not quality management.
    (c) Six Sigma projects are very short term.
    (d) Six Sigma uses advanced statistical tools.

63. Which tool is best for finding multiple causes of a single defect?
    (a) The Ishikawa diagram
    (b) The Pareto chart
    (c) The process flow diagram
    (d) The control chart

64. You are implementing a TQM program. You can choose where to start and what to do first. Historically, the most commonly used order was
    (a) customer delight, internal improvements, work with vendors.
    (b) internal improvements, customer delight, work with vendors.
    (c) internal improvements, work with vendors, customer delight.
    (d) work with vendors, internal improvements, customer delight.

65. Consulting expertise will help with ISO registration in all of these ways *except*
    (a) using an expert reduces cost.
    (b) an expert will give you templates for most required documents.
    (c) the expert will provide training for your entire staff.
    (d) the expert will perform the compliance audit at a lower cost than a registrar.

66. Inherent momentum
    (a) is most beneficial in a kaizen environment.
    (b) is most beneficial in a highly innovative environment.
    (c) in businesses is best resolved by force, as it is with physical objects.
    (d) is a form of negative resistance.

67. Gembutsu are all of these *except*
    (a) things in gemba, such as the defective product.
    (b) to be inspected promptly when a manager arrives.
    (c) things in gemba, including physical equipment and physical evidence.
    (d) essential to process flow.

68. Systems Engineering is all of these *except*
    (a) required in ISO 9000.
    (b) about total systems, which may or may not contain software.
    (c) addresses issues of customer requirements and constraints.
    (d) part of CMMI.

69. We can measure the value of an internal audit department by bench-marking
    (a) the value of clear, factual research results we can use to improve our business.
    (b) the number of audits completed per year, given the size of the department.
    (c) the number of findings issued per year.
    (d) the degree to which the major issues in the risk footprint are covered each year.

70. You can reduce cost of quality the most by
    (a) designing products that can be built with very low defect rates.
    (b) thorough inspection and testing of each product, so that you almost never have to pay for errors found by a customer.
    (c) reducing cycle time.
    (d) using innovative methods and new materials.

71. You run a small manufacturing company. Which of these is *not* a valid reason to avoid Six Sigma?
    (a) Six Sigma only applies if you are making millions or billions of identical products.
    (b) Six Sigma has high training costs.
    (c) Lack of standardization of Six Sigma makes it harder to know if the black belt you hire can do the job at your company.
    (d) Six Sigma will not do much to support continuous improvement.

72. Documentation
    (a) includes specifications, but not records.
    (b) is only used in quality planning and product development, not in QC or QA.
    (c) was invented by creators of the ISO 9000 standard.
    (d) is a cornerstone of quality management.

73. The statement that "In quality management, we've solved the problem in theory but not in practice," points out all of these facts, *except*
    (a) Taylor identified, and for some of his customers, resolved, most quality management problems at the conceptual level by 1911.
    (b) if our quality management program isn't working, we can probably find out why by comparing it to a fundamental text, such as Taylor's *Scientific Management,* or Deming's *14 Points*.

(c) many of the quality management initiatives have worked out in practice solutions to problems Taylor identified in 1911.

(d) we should not have to spend much time on quality management, because all of the problems have already been solved.

74. Objectivity is crucial to improvement. This particular fact is most broadly implemented by which of these quality management tools?
    (a) Fact-based decision-making
    (b) The elimination of slogans
    (c) The cost of quality
    (d) Six Sigma

75. Inspection
    (a) is a form of review.
    (b) requires measurement.
    (c) requires a standard to compare to.
    (d) has been replaced by statistical quality control.

76. If your process uses my outputs, then you are my internal customer. This idea helps reduce all of these problems *except*
    (a) conflict on teams.
    (b) waste in the SIPOC chain.
    (c) poor vertical communication to managers.
    (d) conflict with vendors.

77. All of the following are true about a judgmental sample, *except*
    (a) judgmental sample is an auditing term for a sample based on expert judgment.
    (b) a judgmental sample can be combined with a random sample to create a stratified sample.
    (c) judgmental samples are worse than convenience samples.
    (d) judgmental samples generally have a problem due to self-selection.

78. Software Engineering
    (a) is the CMMI module for software development—writing computer programs.
    (b) is another term for Systems Engineering.
    (c) is required by ISO 9000.
    (d) is used in Six Sigma.

79. ISO 9000 registration is roughly equivalent to all of these, *except*
    (a) the first year of a TQM deployment program.
    (b) a one-year Six Sigma initiative.
    (c) reaching CMM level three, except it is not specific to software engineering.
    (d) running a well-organized, well-documented company.

80. The main reason that auditors do not take responsibility for solving problems is
    (a) auditors do not have the time, they need to focus on auditing.
    (b) auditors are not qualified to solve problems. They are trained as investigators, not implementers.
    (c) if they work on solving the problem, it can compromise their independence.
    (d) management should be made to clean up its own mistakes.

81. A temporary countermeasure is most likely to be used
    (a) to ensure productivity to meet ISO 9000 requirements.
    (b) by a Six Sigma engineer.
    (c) to keep production running while we examine gembutsu.
    (d) at CMMI level 3 or higher.

82. Resistance is
    (a) always unhealthy.
    (b) always healthy, so we need to understand it.
    (c) often, but not always healthy.
    (d) a barrier to be overcome.

83. You are a top executive at a firm that has embraced Six Sigma. The process is failing to show results. Which of the following choices would *not* be likely to help?
    (a) Provide executive leadership, even though Six Sigma formally requires only executive support.
    (b) Evaluate your Six Sigma effort against Deming's 14 points of TQM.
    (c) Provide support for continuous improvement to maintain improvements achieved by Six Sigma projects.
    (d) Shorten Six Sigma project cycle time.

84. A risk footprint is a tool primarily used by
    (a) quality engineers.
    (b) internal auditors.
    (c) external auditors.
    (d) sasquatch hunters.

85. Which of these best describes the relationship between quality engineering and vendor communications?
    (a) In quality engineering, we should begin with vendor communication, because vendors, as suppliers, are the beginning of the SIPOC chain.
    (b) Quality engineering is an internal effort unrelated to vendor communications.
    (c) Quality engineering includes vendor communications about product quality, but we shouldn't discuss the vendor's internal processes.
    (d) Quality engineering should at first focus on internal improvements, then move up the supply chain through vendor communication.

86. All of these are true of workflow optimization *except*
    (a) it is part of lean manufacturing.
    (b) it was developed in lean manufacturing, and has been adopted by Six Sigma.
    (c) it applies primarily to manufacturing.
    (d) it is a made-up term.

87. Forensics means
    (a) related to, or of a standard acceptable to, courts of law.
    (b) the application of science to investigate crimes or business problems.
    (c) any investigation done after the fact.
    (d) related to external auditing.

88. Which is most true of stratified samples?
    (a) Stratified samples are hard to build, but can be very valuable.
    (b) Stratified samples are better than comprehensive samples.
    (c) Stratified samples are less expensive than random samples.
    (d) The term stratified samples is a made-up term.

89. Which of these industries or areas is *not* heavily involved with ISO 9000?
    (a) Europe
    (b) Aerospace

(c) Software development

(d) American automobile manufacture

90. Which of the following would *not* be useful preparation for a poorly managed company that wants to implement Six Sigma?
(a) Achieving ISO 9000 certification.
(b) Achieving CMMI level 3 or its equivalent in your industry.
(c) Focusing on hassle-free management.
(d) Any of a, b, or c would be a good idea.

91. All of these are true about Integrated Product and Process Development *except*
(a) it addresses communication and coordination among stakeholders.
(b) it is used with other components of CMMI.
(c) it helps define software development processes.
(d) it is part of ISO 9000.

92. Fill in the blank. Risk management is to _____, as quality management is to error.
(a) uncertain, possible future events
(b) defined risk
(c) disaster
(d) fiduciary responsibility

93. Which of these would be a good reason to use ISO 9000 rather than TQM?
(a) Your company has a high staff turnover rate, so it is easier to audit than to provide extensive training.
(b) You want to win the Baldridge award.
(c) You don't need to use statistical quality control.
(d) You are preparing to do business in China.

94. Supplier Sourcing
(a) addresses how to ensure that suppliers provide what is needed for your development projects.
(b) is part of Six Sigma.
(c) is part of ISO 9000.
(d) is part of TQM.

95. Muda—the Japanese word for waste, has a very wide meaning. Muda includes all of these, *except*
    (a) overproduction.
    (b) inventory.
    (c) motion.
    (d) failure to use innovative methods.

96. All of these are true about Theory Y, *except*
    (a) Theory Y is crucial in TQM.
    (b) in an environment where Theory Y is operating, managers are more likely to coach than to provide big bonuses.
    (c) in an environment where Theory Y is operating, workers will be likely to have the opportunity to improve.
    (d) if a Theory X manager is replaced by a Theory Y manager, immediate gains in efficiency are the most likely result.

97. Which of the following best describes the effect of continuous improvement—kaizen—on our attitude towards standards?
    (a) Without continuous improvement, we often struggle to meet standards. With continuous improvement, the standard is the minimum requirement for a process, and we strive to improve on it.
    (b) Continuous improvement changes standards through rapid innovation.
    (c) Continuous improvement works better without written standards, as paperwork interferes with the continuous improvement process.
    (d) Continuous improvement changes standards, but does not change inputs.

98. We want to avoid getting a self-selected sample, but that is hard to do in
    (a) selecting products for testing.
    (b) performing customer satisfaction surveys.
    (c) performing employee evaluations.
    (d) analyzing processes.

99. All of these are true of the process of quality definition for our product *except*
    (a) quality definition comes before the definition of processes.
    (b) quality definition is a defined part of ISO 9000.

(c) some quality planning is done before quality definition, but not all of it.

(d) increases the quality of our specification.

100. Six Sigma statistical processes would be most useful at
    (a) CMMI level 1, to accelerate development to higher levels.
    (b) CMMI level 2, after some departments are organized, to help others do as well.
    (c) CMMI level 3, to help maintain quality levels.
    (d) CMMI levels 4 and 5, to improve the ability to deliver high-quality software.

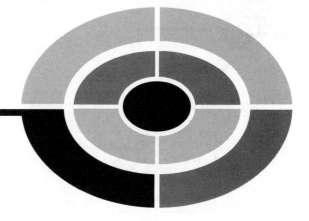

Answers to Quiz, Mid-Term Exam, and Final Exam Questions

CHAPTER 1

1. b   2. c   3. d   4. a   5. c

CHAPTER 2

Year	Person, Organization, or Method	Idea to Match
1755	John Smeaton	Applied the scientific method to engineering problems
1788	James Watt	The centrifugal governor
1788	James Watt	The first feedback device
Around 1850	The US military	Defined the first industry standard
1876	Thomas Alva Edison	The R&D laboratory for new inventions and viable commercial versions of products
Around 1900	Assembly lines	Inspection, followed by rework or discard
1911	Frederick Winslow Taylor	Scientific Management
1920s–1950s	Walter Shewhart	Statistical Quality Control
1920s–1980s	Walter Shewhart and W. Edwards Deming	Plan, Do, Check, Act
1940s–1980s	W. Edwards Deming	Total Quality Management
1947–present	Operations Research	Use of mathematics to improve business decisions
1970	Xerox Palo Alto Research Center (PARC)	The R&D laboratory for maintenance of innovative leadership

CHAPTER 3

1. c   2. a   3. d   4. c   5. c

CHAPTER 4

1. c   2. a   3. b   4. a   5. b

CHAPTER 5

1. c   2. b   3. c   4. e   5. d

CHAPTER 6

1. c   2. a   3. e   4. b   5. c

CHAPTER 7

1. c   2. c   3. a   4. b   5. d

CHAPTER 8

1. c   2. b   3. a   4. a   5. d

CHAPTER 9

1. c   2. c   3. d   4. a   5. b

CHAPTER 10

1. c   2. a   3. d   4. b   5. c

MID-TERM EXAM

1. c	2. a	3. a	4. c	5. a	6. a	7. d
8. b	9. d	10. b	11. a	12. a	13. d	14. b
15. d	16. a	17. c	18. a	19. d	20. b	21. b
22. b	23. d	24. b	25. a	26. d	27. a	28. c
29. b	30. c	31. c	32. a	33. a	34. b	35. b
36. c	37. c	38. a	39. b	40. d	41. d	42. a
43. a	44. a	45. b	46. c	47. c	48. d	49. c
50. c	51. b	52. a	53. a	54. b	55. b	56. b
57. d	58. c	59. c	60. a	61. b	62. a	63. b
64. b	65. c	66. b	67. c	68. d	69. a	70. d

CHAPTER 11

1. b   2. c   3. c   4. b   5. a

CHAPTER 12

1. a   2. d   3. c   4. c   5. a

CHAPTER 13

1. d   2. c   3. b   4. a   5. b

CHAPTER 14

1. b   2. d   3. c   4. a   5. d

CHAPTER 15

1. c   2. b   3. d   4. a   5. c

CHAPTER 16

1. a   2. c   3. d   4. c   5. b

CHAPTER 17

1. a   2. d   3. c   4. b   5. d

CHAPTER 18

1. c   2. a   3. a   4. b   5. d

CHAPTER 19

E-mail your thoughts to *sid@qualitytechnology.com.*

FINAL EXAM

1. a	2. b	3. c	4. d	5. d	6. b	7. a	8. d	9. a	10. a
11. c	12. d	13. d	14. c	15. a	16. c	17. c	18. b	19. d	20. d
21. a	22. c	23. a	24. d	25. c	26. a	27. a	28. d	29. b	30. d
31. d	32. d	33. b	34. c	35. a	36. d	37. d	38. c	39. b	40. d
41. a	42. d	43. c	44. c	45. a	46. a	47. a	48. b	49. b	50. c
51. a	52. c	53. a	54. a	55. a	56. d	57. b	58. a	59. d	60. d
61. d	62. a	63. a	64. c	65. d	66. a	67. d	68. a	69. a	70. a
71. a	72. d	73. d	74. a	75. c	76. c	77. c	78. a	79. b	80. c
81. c	82. c	83. d	84. b	85. d	86. d	87. a	88. a	89. c	90. d
91. d	92. a	93. a	94. a	95. d	96. d	97. a	98. b	99. b	100. d

List of Acronyms and Glossary

Acronyms

Acronym	Description
CWQC	Company-Wide Quality Management
DMAIC	Define, Measure, Analyze, Improve, Control
JIT	Just-in-Time
PDCA	Plan, Do, Check, Act
QA	Quality Assurance
QC	Quality Control
QP	Quality Planning
SOP	Standard Operating Procedure
TQC	Total Quality Control
TQM	Total Quality Management
VOC	The Voice of the Customer

Glossary

Accurate	A measurement is sufficiently accurate if it can measure closely enough to ensure that the attribute is within tolerances.
Architecture	The process of looking at something from several perspectives, focusing on quality within constraints from multiple perspectives.
Benchmarking	A defined measure of productivity based on comparison to similar processes.
Best practice	The best way to do a repetitive process at your organization, that is defined, repeatable, written down and derived from industry best practices.
Buddy programming	An innovative approach that actually has two people working simultaneously at one computer screen to write code.
Causal loops	A cause and effect pattern where the effect makes the cause more likely, creating a repeating, increasing, or decreasing cycle.
Cause	The reason for the good or poor performance.
Cause-and-effect diagrams	Ishikawa diagrams.
Cause-effect pair	A unique pair of events that describe a cause that results in a defect of product or service.
Change control	The management of changes to requirements.
Checking	A general term we will use for any activity that compares the product or service against requirements, or that compares the technical process of creating the product or service against requirements.
Company-wide quality control	A Japanese term for Total Quality Management.
Compare	To check for similarities, differences, or conformance.
Comprehensive sample	A sample as close to the entire population as possible. The difference between a population and a comprehensive sample is due to mistakes such as missed items or lost data.
Condition	The measured or assessed level of performance of how things are or were done.
Conformance	The matching of an attribute to its requirement.

Consequence	What would happen if an error went undetected.
Continuous improvement	An incremental approach to quality management involving all employees. Contrast innovation.
Convenience sample	A sample gathered in the easiest way possible.
Corrective action	A response to a defect, such as rework or scrapping. Also, an immediate response to a process failure, such as getting a machine running again.
Criteria	What should or could be—for example, the relevant standard or best practice.
Croque monsieur	A high quality ham-and-cheese sandwich. For precise specifications, see *www.qualitytechnology.com/QMD*.
Data flow diagramming	A structured technique for mapping the flow of data.
Defect	An input, output, or product attribute that is outside tolerances.
Design of experiments	The process of designing an experiment to give us useful results for quality management.
Design	A field of work that looks for effective and efficient solutions that involve engineering, aesthetics, and market preferences.
Destruct tests	Tests that check a feature, but destroy the product in doing so.
Document	An ISO 9000 term for an SOP.
Effect	The gap between condition and criteria, and its consequences.
Effectiveness	Ability to achieve the desired effect.
Efficiency	Ability to do a process at minimal cost or using the minimum of particular resources.
Egoless programming	The idea that, in programming, we need to do our best work, but then accept correction from the team, and the team needs to correct the work without commenting on the worker.
Equipment	Tools.
Error	A process attribute that is outside tolerances.
Estimation	Always means making statements about the population based on statistics from the sample.
Event	Opportunity.
Examine	To look at.

Experimental control	A group that is not changed, the results of which are used for comparison with a group that has an intervention applied.
Exposure	Vulnerability to lawsuits, criminal charges, loss of reputation, or other costs.
Fiduciary risks	Risks related to our responsibility to others outside our company.
Finding	A statement of a material difference between conditions and standards.
Fishbone diagrams	Ishikawa diagrams.
Five S's of good housekeeping	Sort, Straighten, Scrub, Systematize, and Standardize.
Flow diagramming	A structured technique for mapping processes and decisions.
Flow production	Where workstations are arranged in order of work flow.
Forecasting	Always means predicting expected future measures or results based on past measures or results.
Forensic	Of a standard acceptable for, or related to, a court of law.
Gemba kaizen	The Japanese quality management process that includes continuous improvement with a focus on small changes to the place work actually happens.
Gemba	The place where the company makes a difference for the customer or meets the customer. In manufacturing, the shop floor. In service industries, any location where an employee might meet a customer.
Gembutsu	Things in gemba—physical equipment and physical evidence, such as the defective product.
Impact finding	An auditor's statement as to the consequences of findings.
Inherent conservatism	All lasting systems' tendency towards homeostasis—a tendency to forcefully move towards their own pre-defined point of balance.
Inherent momentum	Systems, like physical objects, tend to keep moving in the same direction. A successful change of direction requires not force, but encouragement for habit change.

Innovation	An approach to quality management that focuses on breakthroughs and leaps forward. Contrast Continuous Improvement.
Input requirement	A requirement for an input, such as a raw material or a subcomponent, provided by a vendor, before using it in our own process. Or a requirement for an input to any process at any level.
Inputs	The ingredients, raw materials, or components that go into a process and become part of the output.
Inspection	The act of examining an attribute of an input, process, or output to compare it to a requirement, with the purpose of identifying defects.
Integrated product and process development	A CMMI component that addresses communication and coordination among stakeholders, and is used in conjunction with one or more other disciplines.
Intervention	An act in an experiment, of which we want to see the results.
Ishikawa Diagrams	A diagramming methods that helps identify multiple causes of a single error or defect.
Iteration	Repeating at more and more refined levels.
Judgmental sample	An auditing term. A sample created using our own common sense—our expert judgment—to decide how to take our sample.
Just-in-Time manufacturing	The approach to manufacturing, developed by Toyota, where items are produced only to fill orders placed by the customers, and waste (muda) in the production stream is minimized. It only works in an environment already optimized by gemba kaizen.
Kaizen	Continuous improvement, both in general and as a Japanese quality management method.
Lean manufacturing	Just-in-Time manufacturing.
Liability	Potential cost that we will, or might, have to pay in the future.
Management response	A statement by management in response to an auditor's finding, and included in the audit report, that describes how management will resolve the problem.

Muda	The Japanese word for waste, has a very wide meaning. Muda includes all items and activities that do not add value to the final customer product, including: overproduction, inventory, rejects, motion, processing, waiting, transport, and time.
Non-conformance	An error or defect.
Objective	As a quality of checking, giving the same results no matter who does the checking.
Opportunity	A moment in a process where an error can occur, or an incident in a product where a defect can be present.
Output requirement	A definable, measurable feature of the output of an entire process or a sub-process (a product or component).
Outputs	The end results of a task, such as a component or a finished product.
Pareto optimization	A process that prioritizes defects for the prevention effort.
Permanent preventative solution	Solution to a root cause, so that this error and ones like it will not happen again in this process or any similar process within the company.
Plan, do, check, act	The cyclical process of correcting problems created by Shewhart, promoted by Deming, and applied to all gaps and errors by kaizen.
Planned obsolescence	The intentional creation of low quality to drive up billable repair costs.
Population	The entire range of items in the world that we want to know about through statistical inquiry.
Process	The activity of changing inputs or outputs— the work.
Process requirement	A requirement on a defined measure of a process measured while it is happening.
Pull production	Where each workstation makes only enough output to supply input as needed for the next step.
Purchaser	Person who decides to buy and pays for a product or service.
Quality assurance	QA. Any and all activities—especially activities that cut across departments—that we do to make sure that we do our quality work right.

Quality assurance	Includes activities to evaluate and improve processes, re-engineer work to eliminate unnecessary processes or steps, ensure effective communication and mutual understanding throughout the SIPOC chain, and auditing and review to ensure all processes are maintained to standard and improved.
	Includes quality activities outside the realm of checking and quality control. QA includes cross-departmental communication about quality, communication with vendors, redesign of the product or process to prevent error, and a variety of audit processes to make sure that work and management are being done to standards or in accordance with best practices.
Quality control	Two meanings. The broader meaning is synonymous with checking; the narrower meaning is Statistical Quality Control.
	In the broad sense including all forms of checking, ensures that outputs and processes meet requirements, that defective output is reworked or scrapped, and that all seven aspects of processes are adjusted and restored to work within tolerances.
	The narrow meaning refers to statistical quality control, where we test a small sample of the entire product batch and extrapolate to define qualities of the entire batch. In the broader use, quality control is synonymous with checking. It refers to all activities of review, inspection, and testing of the product or its technical process, with or without sampling and statistics.
Quality control plan	A requirements tracing matrix expanded to include our quality control plan, showing how each requirement is associated with various checks and tests.
Quality definition	The definition of all elements of our product that add value for the customer or stakeholders, or are required for our product or service to meet relevant standards and regulations.

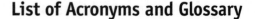

Quality planning	Includes all early efforts to plan how error will be prevented (QC) and how quality will be managed (QA), and some of the design activities of previously part of QA. The planning that includes defining what processes are required to deliver the product to meet or exceed specifications, putting them in order by linking outputs of one process to inputs of the next, and then defining all seven aspects of each process with requirements and tolerances on all key variables, so that we can consistently produce all outputs of all processes to specification.
Quality	That which adds value. To the extent that our specification is error-free and accurately represents what the customer wants and we deliver a result meeting or exceeding that specification, we deliver quality.
Quota sample	The quota sample is similar to the convenience sample, except that we stop collecting when our sample is large enough.
Random sample	A sample in which each item in the population has an equal chance of being included in the sample.
Record	An ISO 9000 term for a checklist, quality control results, or other document that shows work on a particular process was done at a particular time according to the document for the process.
Registrar	An independent company that, for a fee, audits you to see if you meet—and later to see if you are maintaining—compliance with the ISO 9000 standards. Each registrar is, in turn, registered with an agency—a different one in each country around the world—that ensures all registrars are auditing independently and properly ensuring that they only issue certificates to companies that truly meet the ISO 9000 standard.
Requirement	A defined quality of an input, process, or output that is necessary for quality.

Requirements definition	The process of taking all of the requirements from different sources and combining them into a requirements specification.
Requirements elicitation	The process of dialog with a customer to define customer requirements.
Requirements specification	A document including all requirements for a product, service, or process.
Requirements tracing matrix	A document that links each requirement to its source (customer, stakeholder, or standard), and to the features of each component required to achieve the customer requirement.
Resources	Those things including disposable items (such as cleaning supplies) and our effort, which are used up in the process but do not get included in the product.
Review	The process of comparing a document, such as a requirements specification or a design plan to standards or requirements that govern the process or results required of that item. Reviews can be highly formal and strict, or they can be loose and informal.
Risk footprint	A prioritized diagram of risks prepared by auditors in their annual planning.
Root cause	The deepest, most basic reason for an error or defect.
Sample	A selected part of a population.
Scientific management	The predecessor of all quality management, using the scientific method to define, then continuously improve best tools and methods for doing any job.
Scope	What we are making, and all the work we will do to make it.
Self-selected sample	In this case, the subject has a say in whether or not to be included in the sample. A good example of a self-selected sample is customers who choose to answer our customer survey. Unfortunately, we can't be sure that the self-selection doesn't bias the sample.

Six Sigma	An approach to project management developed in North America in the 1990s that emphasizes breakthrough innovation.
Software engineering	Engineering for software development—writing computer programs.
Software inspection	TQM applied to software development.
Standard operating procedure	A document that tells how to do a process.
Statistical process control	Statistical Quality Control.
Statistical quality control	The application of sampling and statistics to inspection.
Stratified sample	Building a stratified sample is complicated, but essentially, a stratified sample is a combination of a judgmental sample, then random items selected from groups selected on the basis of expert judgment.
Stretch target	The high end of the range of a goal set by executives to foster innovation.
Subjective	As a quality of checking, giving different results depending on who does the checking
Supplier sourcing	A CMMI component that addresses how to ensure that suppliers provide what is needed for your development projects.
Synergy	The quality of people being able to do more together than we could do separately.
Systematic sample	A sample gathered in a patterned, non-random way.
Systems engineering	Engineering of total systems, which may or may not contain software. It addresses issues of customer requirements and constraints.
Techniques	The instructions for the work process.
Testing	The process of actually doing something with a product, service, or component and seeing what happens.
Tolerance	The acceptable range of variation around a desired goal.
Tools	Those things which are used for the task, but not used up.
Total quality control	A Japanese term for Total Quality Management.
Total Quality Management (TQM)	The method of quality management developed in Japan by W. Edwards Deming and Japanese companies.

Unbiased	Free of an error that consistently throws off the results in one direction or the other.
Unhealthy resistance	The results of poor communication and lack of willingness to create good communication and eliminate misunderstanding. This is covered by the word politics in the negative sense of the term, and includes the effects of all forms of dishonesty and avoidance.
User	Person who actually uses a product or service.
Voice of the customer	A catch phrase to remind us that we need to be certain we are addressing customer requirements, not purely internal requirements or our idea of what the customer wants.
Work breakdown structuring	A tool from project management used to define all the parts of a product or service or result, and all the work—and only the work—necessary to create it.
Work environment	The surrounding space around the work, with the attributes required for the work to succeed.
Workflow optimization	The process of organizing a production line to minimize wasted time.

Resources
for Learning

There is always more to learn about quality management. For a complete, annotated bibliography and list of resources on the Internet, visit *www.qualitytechnology.com/QMD*. Please send any questions or thoughts to *sid@qualitytechnology.com*. I am always delighted to hear from my readers and help in whatever way I can.

Bibliography

Arter, Dennis R, and J. P. Russell. *ISO Lesson Guide 2000 Pocket Guide to Q9001–2000*. Milwaukee, WI: ASQ Quality Press, 2001.

Bateson, Gregory. *Mind and Nature a Necessary Unity*. New York: Dutton, 1979.

Bethune, Gordon, and Scott Huler. *From Worst to First Behind the Scenes of Continental's Remarkable Comeback*. New York: Wiley, 1998.

Bogan, Christopher E. and Michael J. English. *Benchmarking for Best Practices Winning Through Innovative Adaptation*. New York: McGraw-Hill, 1994.

Brassard, Michael, Diane Ritter, and GOAL/QPC. *The Memory Jogger II a Pocket Guide of Tools for Continuous Improvement & Effective Planning.* Methuen, MA: GOAL/QPC, 1994.

Brue, Greg. *Six Sigma for Managers.* A Briefcase Book. New York London: McGraw-Hill, 2002.

Crosby, Philip B. *Quality Is Free the Art of Making Quality Certain.* Mentor, 1979.

Crosby, Philip B. *Quality Without Tears the Art of Hassle-Free Management.* New York: McGraw-Hill, 1984.

Dobyns, Lloyd and Clare Crawford-Mason. *Thinking About Quality Progress, Wisdom, and Deming Philosophy.* New York: Times Books/Random House, 1994.

Gale, Bradley T. and Robert Chapman. Wood. *Managing Customer Value Creating Quality and Service That Customers Can See.* New York: Free Press, 1994.

Garvin, David A. *Managing Quality the Strategic and Competitive Edge.* New York London: Free Press Collier Macmillan, 1988.

Goldratt, Eliyahu M. *The Goal the Process of Ongoing Improvement.* North River Press, 1984.

Hickman, Craig R. and Michael A. Silva. *Creating Excellence Managing Corporate Culture, Strategy, and Change in the New Age.* New York: New American Library, 1984.

Howe, Roger J., Dee Gaeddert, and Maynard Alfred. Howe. *Quality on Trial Bringing Bottom-Line Accountability to the Quality Effort.* New York: McGraw-Hill, 1995.

IIL. *Advanced Project Quality Management,* 2005.

Imai, Masaaki. *Kaizen the Key to Japan's Competitive Success.* Random House, 1986.

Imai, Masaaki. *Gemba Kaizen a Commonsense, Low-Cost Approach to Management.* McGraw-Hill, 1997.

Keller, Paul. *Six Sigma Demystified.* McGraw-Hill, 2005.

Lane, Tom. *The Way of Quality.* Bard Productions, 1994.

Lickson, Jeffrey. *The Continuously Improving Self a Personal Guide to TQM.* Fifty-Minute Series. Los Altos, Calif.: Crisp Publications, 1992.

Phillips, Ann W. *ISO 9001:2000 Internal Audits Made Easy Tools, Techniques, and Step-by-Step Guidelines for Successful Internal Audits.* Huntsville, AL: Quality Techniques, 2002.

Roberts, Harry V. and Bernard F. Sergesketter. *Quality Is Personal a Foundation for Total Quality Management.* New York Toronto New York: Free Press Maxwell Macmillan Canada Maxwell Macmillan International, 1993.

Russell, J. P. *The Internal Auditing Pocket Guide Preparing, Performing, and Reporting*. Milwaukee, Wisconsin: ASQ Quality Press, 2002.

Shelton, Ken. *In Search of Quality*. Provo, UT: Executive Excellence Pub., 1996.

Townsend, Patrick L. and Joan E. Gebhardt. *Quality in Action 93 Lessons in Leadership, Participation, and Measurement*. New York, NY: J. Wiley, 1992.

Walton, Mary. *The Deming Management Method*. New York, NY: Putnam, 1988.

INDEX

INDEX

ABOUT THE AUTHOR

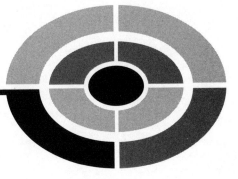

Sid Kemp, a certified project management professional (PMP), is a nationally recognized consultant and trainer. His company assists Fortune 500 companies, major governmental agencies, and small businesses in quality management, strategic planning, process improvement for project management and operations, improvement of audit methodologies, and deploying new technologies. Sid is also the author of eight business books, including *Project Management Demystified.*

Sid is delighted to hear from his readers and answer their questions. You can reach him, and also learn more about his training, consulting, and coaching services, at www.qualitytechnology.com.